ENERGY CONSERVATION TECHNIQUES FOR ENGINEERS

ENERGY CONSERVATION TECHNIQUES FOR ENGINEERS

Harry B. Zackrison, Jr., P.E.

VNR VAN NOSTRAND REINHOLD COMPANY
NEW YORK CINCINNATI TORONTO LONDON MELBOURNE

Library of Congress Catalog Card Number: 82-23691
ISBN: 0-442-29392-5

Manufactured in the United States of America

Published by Van Nostrand Reinhold Company Inc.
135 West 50th Street
New York, New York 10020

Van Nostrand Reinhold Company Limited
Molly Millars Lane
Wokingham, Berkshire RG11 2PY, England

Van Nostrand Reinhold
480 Latrobe Street
Melbourne, Victoria 3000, Australia

Macmillan of Canada
Division of Gage Publishing Limited
164 Commander Boulevard
Agincourt, Ontario M1S 3C7, Canada

15 14 13 12 11 10 9 8 7 6 5 4 3 2 1

Library of Congress Cataloging in Publication Data
Zackrison, Harry B.
Energy conservation techniques

Includes index.
1. Energy conservation. 2. Buildings—Energy conservation. 3. Power resources. I. Title.
TJ163.3.Z3 1983 696 82-23691
ISBN 0-442-29392-5

PREFACE

Throughout this book the theme is woven for understanding the very real difference between energy conservation and energy management. In general wherever the term energy conservation is utilized it can be construed to mean and imply the use of the concept of energy management. Regardless of whether the energy savings are based on lighting or power reduction, the concern must be in elevating lost productivity so as not to increase the hours of operation required to produce the same goods and services which would lead to increased total energy use. We must be careful in our design analysis so that energy savings in one discipline do not increase energy consumption in another because of improper evaluation and application, thereby causing that facility's design to be much more energy intensive (consume more energy needlessly).

In one simple handy book all of the major, most important, cost-effective, energy-saving information is readily available for reference and application by both professional engineers and architects as well as laymen. It has potential value to electrical and mechanical engineers, plumbing engineers and designers as well as, to a somewhat lesser degree, the residential homeowner or condominium home or office owner. I believe this to be the first book of its kind to completely cover energy management and conservation of both lighting and power between the covers of one not overly large book and that it is especially unique in its presentation of innovative energy conservative approaches.

In view of the high cost of energy brought on by the 1973 Arab oil embargo and the subsequent Iranian crisis, every facet of American life is affected. The Department of Energy (DOE), NASA, NAVFAC, GSA and others have proliferated tons of paper on various forms of proposed or mandated energy reforms. Each state has its own energy code that is mandated by the federal government in the form of sanctions it can place and the federal grants and expenditures it can withhold from each state. Since all practicing engineers and architects must adhere to these local, state and federal codes, it behooves them to have the most complete, simplified and easy-to-understand energy conservation methodology readily at hand.

Every household is faced with high interest rates and high costs of living, with prices for clothing, food, gasoline, oil, gas and electricity, to name but a few, going out of sight. As a result of these current events, this book can help alleviate the crunch of some of those energy related costs to both homeowners, and owners of large office buildings and facilities.

But just as important to owner and building operators is that the concepts developed throughout this book can be money-saving approaches, reducing both first costs and life cycle costs. This, therefore, becomes an economic inducement for owners and builders to comply with meaningful energy and resource conservation codes and to utilize the services of experienced and competent engineers knowledgeable in energy conservation. These conservation techniques will be beneficial to the country in terms of energy and resources conserved and to its people in terms of money saved during these inflationary times.

Specifically, the items outlined apply equally well to retrofitted structures as to new construction. In many cases, the additional first cost of renovation or retrofit can be paid back in less than one year from these resource-conserving and energy-saving techniques. For retrofit projects the cost of renovation not only saves on resources, energy and fuels as well as reduces fuel and electric bills, but the accrued savings pay back the additional cost for the renovated structures in terms of both principal and interest.

For their help in editing, organizing and in general insuring the technical quality and correctness of the enclosed material I wish to extend my thanks to the following: O.J. Vogl of the Society of American Value Engineers and Hughes Aircraft Company; Donald C. Thomas, Manager of Commercial Engineering, U.S. Lighting Division of GTE/Sylvania; Robert Bonazoli, Director of Engineering, U.S. Lighting Division of GTE/Sylvania; Mr. John Sainz, Sales Manager, Power Capacitors Industrial Markets, General Electric Company; Mr. John Harder, Advisory Engineer, Westinghouse Electric Corp.,

Distribution Apparatus Division; and Lloyd M. Costley, Esq. This book is dedicated to the purpose of improving the quality of life for mankind, reducing the burden caused by the high cost which was created by the energy crunch and to increasing the chances that our posterity will live to enjoy our present standards of living. It is hoped that although we have a finite quantity of precious resources at our disposal that it and the wonders of our world will still be at the disposal of yet unborn generations to come. May their living standards be of a quality as high if not higher than our own. This book is especially dedicated to my wife, Eileen; daughter, Linda; and two sons, John and Paul. May this book be an inspiration to my children in their formal education and pursuit of vocational excellence and avocational happiness.

HARRY B. ZACKRISON

CONTENTS

ENERGY CONSERVATION TECHNIQUES FOR ENGINEERS

1 INNOVATIVE ENERGY-SAVING ELECTRICAL DESIGN TECHNIQUES

INTRODUCTION

Energy analyses at one time required the consideration of resource utilization factors (RUFs) and resource impact factors (RIFs). These numerical weight factors were each multiplied by a numerical value attributed to various systems, and the smallest product indicated the most energy efficient fuel to use for a particular application. Therefore, each energy budget in (Btu/ft^2)/year was multiplied by a RUF or a RIF which related the budget to raw resource energy. RUFs considered how efficiently one source of energy could be converted into another, i.e., 3.04 Btu of heat energy are required to produce and transmit 1 Btu of electrical energy to the point of use. RIFs took into account the relative availability or scarcity of a fuel —whether it was nonreplenishable, such as oil; or replenishable, such as hydroelectric power.

In the past, consideration was given to source and boundary concepts in building designs; hence the origin of RUFs and RIFs. In the *boundary concept*, energy and materials used were considered only within the confines of the building and outwards 5 feet from the exterior of the building walls. In the *source concept*, the cost in labor, materials, and energy expenditure to produce a unit of any type of energy or a unit of a given material commodity was evaluated. Consideration was also given to the impact that other disciplines have on the final resource utilization factor.

The following narrative and analysis will consider a number of electrical design techniques that contribute to minimizing energy consumption and maximizing the efficiency of electrical designs and installations by applying the source concept of energy expenditure. By first reducing the raw products required in a finished structure and the quantity of finished materials used, the amount of initial energy expenditure to produce the total building will be reduced.

But just as important to owner and building operators is that these are money-saving approaches, reducing both first costs and life-cycle costs. They offer, therefore, an economic inducement for owners and builders to comply with meaningful energy codes and utilize the services of experienced and competent electrical engineers knowledgeable in energy conservation. These energy conservation techniques will benefit the country in terms of energy conserved and its people in terms of money saved during these inflationary times.

DISTRIBUTION UTILIZATION VOLTAGE

As much as possible, a service voltage of 277/480-volts, 3-phase, 4-wire should be utilized for all balanced 3-phase loads such as motors and electrical resistive heating, and for all overhead lighting systems, instead of using what is sometimes considered the more conventional 120/208-volt, 3-phase, 4-wire system. This will achieve a reduction of 57% in current. (See Figure 1–1(A).)

$$\frac{(120\ \text{V} - 277\ \text{V})}{(277\ \text{V})} \times 100 = -57\%$$

For a given wattage the amount of current is directly proportional to the voltage used. Due to its higher load-carrying capacity, 277/480-volt wiring systems can carry 231% as much energy as 120/208-volt system conductors. (See Figure 1–1(B).)

$$\frac{277\ \text{V}}{120\ \text{V}} \times 100 = 231\%.$$

Also, because of the nonlinearity of the current-carrying capacity of copper conductors, larger conductors do not carry proportionally higher current capacities. (See Figures 1–1(A) and 1–1(B).) Therefore, the size and number of current-carrying conductors and raceways are increased by over 60 percent.

By using a higher voltage, the I^2R heat losses will be lower and voltage drop will also be less severe and less of a problem. The equation I^2R might imply that as the ampacity of our system increases and thus increases conductor size, we might expect that the energy line losses in the form of heat to increase as the square of the increase

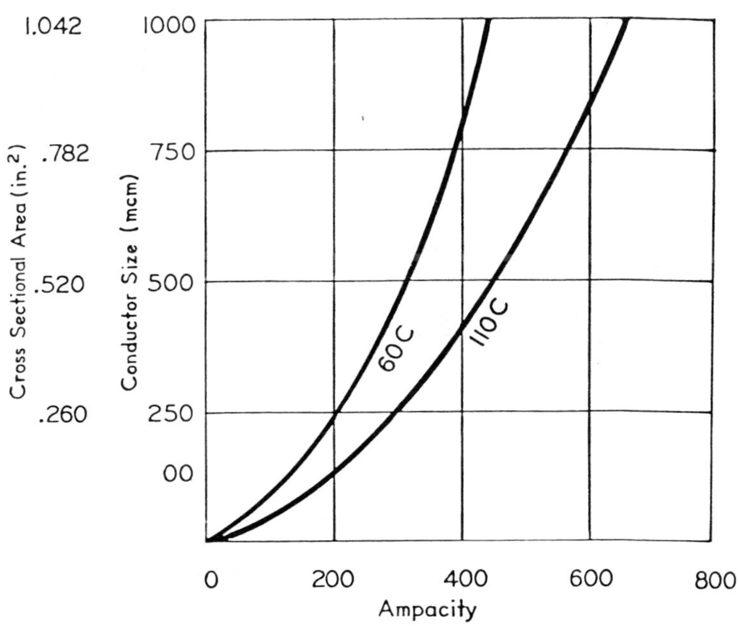

Allowable ampacity for 250 mcm copper

Type	Insulation	Ampacity
TW	60C	215 A
AVA	110C	315 A

Figure 1–1(A). Comparison of conductor insulation and ampacity.

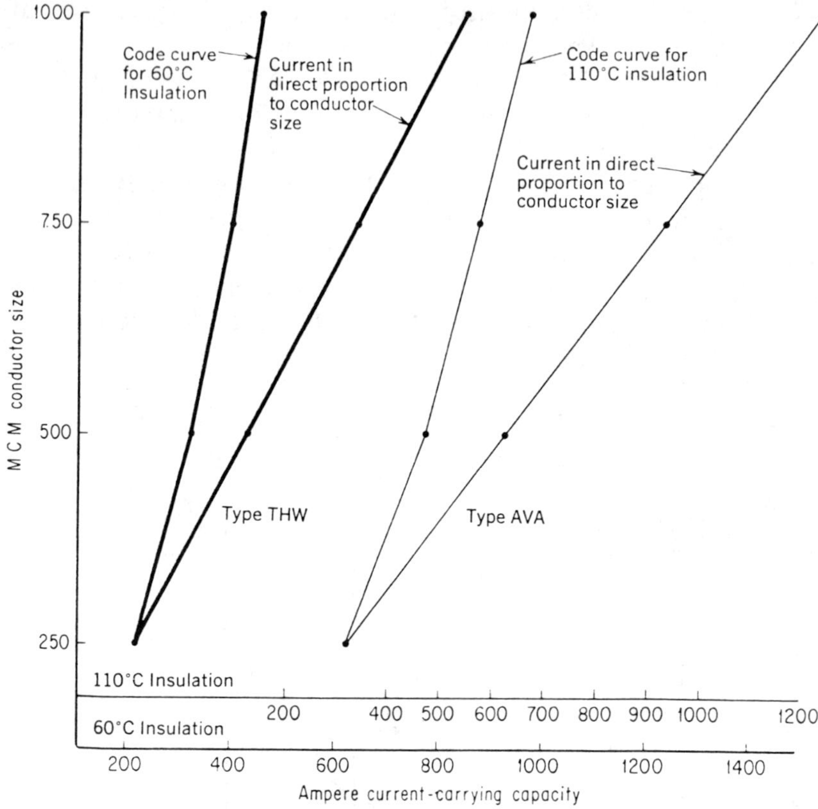

conductor M C M	Current capacity	
	Type 60°C	Type 110°C
250	215 A	315 A
500	320	470
750	400	580
1000	455	680

Figure 1–1(B). Ratio of conductor size amd ampacity.

in current. Fortunately for us this is not the case, since as we previously stated, the current-carrying capacity of conductors does not increase in direct proportion to the increase in conductor size but at a reduced rate. As a result, a conductor of twice the current-carrying capacity has more than twice the area (between 2.5 and 2.65), one and one-half times the diameter, and more than three times the size in circular mils (CM). As a result of this real world fact, resistance (*R*) values are not decreasing in direct proportion to conductor ampacity increases but are decreasing more rapidly. Resistance of a conductor is the product of its resistivity (ρ) times its length (L) divided by its area (A): $R = \rho L/A$. Thus, since the size of the conductors does not increase in direct proportion to ampacity but instead increases at a much more rapid rate, the increase in size of the conductor has its resistance decrease at a greater than linear rate (non-linear rate decrease). Hence, the resistance is decreasing at about three to nearly three and a half times for every doubling of the current capacity, depending on ampacity size and ampacity range of the conductors.

Thus a #2 American Wire Gauge (AWG) conductor, 66,360 circular MIL (CM) size, has a 115 ampere capacity at 0.202 ohm of resistance. A #4/0 conductor, 211,600 CM size, is 3.19 times larger than the #2 AWG and has an exactly doubled ampacity of 230 amps at 0.0640 ohm of resistance* which is precisely 3.44 times lower than the #2 AWG conductor. As a result, a #2 AWG conductor with a full-load current capacity, for a 100-foot length of single conductor, would expend 267.1 watts of energy in wasted heat, $[P = I^2R = (115\ \text{amps})^2 \times (0.0202\ \text{ohm}) = 13{,}225\ \text{amps}^2 \times (0.0202\ \text{ohm})]$. The #4/0 conductor will waste 338.56 watts of energy in the form of heat, $[P = (230\ \text{amps})^2(0.00640\ \text{ohm}) = 52{,}900\ \text{amps}^2 \times 0.00640\ \text{ohm})]$. The ratio of energy wastage in the form of I^2R heat loss is 338.56 watts divided by 267.1 watts or 1.27 to 1—not current squared, $(2I/I)^2$ or 4 to 1.

A #3/0 conductor rated at 200 amps capacity has a resistance value of 0.00805 ohm for a 100-foot length of run and in conjunction with a 200-amp overcurrent protection device is capable of transporting 160 amps continuously. This is based upon the National Electrical Code (NEC) Article 384–16(c), which requires that "the total load on any overcurrent device located in a panelboard shall not exceed 80 percent of its rating where in normal operation the load will continue for 3 hours or more," except "where the assembly including the overcurrent device is approved for continuous duty at 100 percent of its rating." A 500-MCM conductor rated at 380-amp capacity has a resistance value of 0.00294 ohm for a 100-foot length of run and in conjunction with a 400-amp overcurrent protection device is capable of transporting 320 amperes of current continuously as per Article 384–16(c) of the 1981 NEC. The 380-amp capacity, 500-MCM

*Per 1000 feet of length.

conductor is allowed to be used in conjunction with a 400-amp capacity overcurrent control device as per the 1981 National Electrical Code Article 240–3 exception No. 1, which states that "where the ampacity of the conductor does not correspond with the standard ampere rating of a fuse or a circuit breaker without overload trip adjustments above its rating (but which may have other trip or rating adjustments), the next higher standard device rating shall be permitted only if this rating does not exceed 800 amperes." The 500-MCM conductor has 500,000 CMs as compared to the #3/0's 167,800 CMs which is a ratio of 2.97 to 1, but its resistance is less than the #3/0's by a ratio of 2.73 to 1. Although these ampacities are not 2 to 1, by code they each are allowed to be tailored to overcurrent devices that are in a ratio of 2 to 1. As a result, a 500 MCM, at full, overcurrent-rated, continuous-duty, amperage capacity, consumes 301 watts in wasted heat energy per 100 feet, as compared to a #3/0 conductor's 206 watts for a ratio of 1.46 to 1 which is far short of a 4 to 1 ratio based solely on the formula I^2R.

The primary concern to remember is not whether we are increasing the conductor size, since the energy loss will not increase as the square of the current (or four times as much, by doubling the current), but to use the highest voltage possible in order to decrease the current density. In this way large amounts of accumulated smaller energy losses due to conductor line losses will be alleviated not by 4 to 1 but by 1.27–1.46 to 1 ratios.

Voltage drop is also less severe at higher distribution voltages. In very few cases should secondary distribution at 120/208 volts be recommended for a particular project, since the additional first cost of feeders and raceways will be much higher (most radial feeders may also have to be oversized to compensate for voltage drop due to IR losses on the line).

These advantages of a 277/480-volt distribution can, of course, be negated by serving the majority of the lighting or other major loads at 120/208 volts via local step-down transformers. When a large portion of the building load is served at this lower voltage, the local step-down transformers become larger as do their core and coil losses, introducing further energy wastage. To serve a given load, the transformer loss is now incurred twice, once at about 1.85% at the main primary to 277/480-volt transformer, and again at almost 5% in the 480 to 120/208-volt step-down transformers.

It is also possible to minimize the extent and cost of electrical services by deviating from the current trend of using lower voltage distribution design and equipment. In an office building or school facility, 120/208-volt, 3-phase, 4-wire service for receptacles and miscellaneous power should normally account for no more than 5% to 8% of the total load. If the proportion of the 120/208-volt load can be held down to this percentage (many electrical design engineers routinely serve 40% to 50% of the load at 120/208 volts), then it is possible to eliminate the provision for one of two different primary transformers sometimes used in order to produce two different incoming secondary service voltages (by elimination of the 120/208-volt transformer and service entirely and installing only a 277/480-volt transformer and service). The small receptacle and miscellaneous power load can then be fed from small local step-down transformers located at the centers of load throughout the building.

Within the building, power should be distributed generally at 277/480 volts to serve lighting, motor, heating, ventilating and air conditioning (HVAC), and cooling loads. 120/208-volt receptacles and miscellaneous power panels as required should then be subfed, via local step-down transformers, from 277/480-volt lighting and power panels.

PRIMARY TRANSFORMERS

Primary transformers should be a liquid type as opposed to a dry type due to their longer life expectancy—60 to 80 years rather than the 12 to 15 years common with dry-type primary transformers. This represents an additional mining and refining energy savings above and beyond building energy savings by extending the useful life of existing and finished goods. Although the pentachlorobiphenyl or PCB compounds formerly used to cool indoor, liquid-type, medium-voltage (601 to 15,000 volts) primary transformers are now obsolete and increasingly unavailable (they are commonly known by trade names such as Askarel, Pyranol and Inerteen), they are being replaced with new biodegradable compounds. Monsanto Chemical Company of Canada has imposed an importation moratorium on PCBs since October 1976. General Electric has not sold any new transformers supplied with PCBs since March 31, 1977, and Westinghouse has not sold any new transformers filled with PCBs since December 1976. The

Toxic Substances Control Act (TSCA) and the Occupational Safety and Health Act (OSHA) has since banned PCB from use in any new product.

Inert compounds that meet TSCA minimum requirements and are approved by OSHA must be biodegradable, flame-resistant, and have the following chemical characteristics: (1) a highly saturated paraffinic oil whose base fluids are refined by solvent extraction with hydrogenation and processed from high molecular weight paraffinic crudes, and (2) silicone (polydimethylsiloxanes), long chained, high molecular weight polymers consisting of stable SI-O-SI mers. The 1981 NEC does, however, make a distinction between the various fluids on the market. Fluids such as RHEtemp and the high weight paraffinic crudes, listed as merely for High-Fire-Point, Liquid-insulated transformers, are less flammable liquids as per Section 450–23 and are permitted to be installed without a vault in noncombustible occupancy areas of noncombustible buildings, provided there is a liquid confinement area, and the minimum clearances required by the heat release rates of the listed liquid are maintained.

Such indoor transformer installations not meeting the clearance requirements of the liquid listing, or installed in combustible buildings or combustible occupancy areas, shall be provided with an automatic fire extinguishing system or shall be installed in a vault.

In general, silicones, transformer grade tetrachloroethelyene, and freon meet Article 450–24's requirements for nonflammable fluid-insulated transformers. Transformers insulated with a dielectric fluid identified as nonflammable shall be permitted to be installed indoors or outdoors. For the purposes of this section, a nonflammable dielectric fluid is one which does not have a flash point or fire point, and is not flammable in air.

On the other hand, Askarel-insulated transformers installed indoors and rated over 25 kVA must be furnished with a pressure-relief vent and where installed in a poorly ventilated place must be furnished with a means for absorbing any gases generated by arcing inside the case, or a pressure-relief vent shall be connected to a chimney or flue that will carry such gases outside the building. Hence, liquid-type transformers are still a viable choice for primary to secondary step-down voltage applications. As a rule, liquid-filled, primary, medium-voltage, step-down transformers have a full-load core-and-coil loss of less than 2% of their kVA rating at 0.9 power factor. Since medium-voltage transformers are subject to very high impulse levels and require high basic impulse ratings, aluminum coils normally should not be considered over copper even in light of lower weight and lower first cost. Generally, most manufacturers that offer the aluminum winding option in medium-voltage transformers do not supply it for the primary windings due to the high impulse levels. Generally it is the lower-voltage secondary coils that are supplied in aluminum while the primary windings are supplied with copper. Some manufacturers still use a lower tensile strength aluminum versus the newer triple E aluminum annealed grade which is more comparable in strength to copper. However, aluminum has higher internal I^2R core-and-coil losses than copper. In addition, incoming primary-service transformers should not be wired wye-wye because third harmonic currents (generated by fluorescent and HID ballasts) can cause balanced 3-phase circuit neutrals to carry as much as 90% to 95% of the phase currents. If the primary is wired in wye, then this reflected neutral load loss will flow in the primary feeder neutral. Electrical load losses through the primary neutral are an unnecessary waste of energy, leading to higher energy wastage and costs, as well as to greater maintenance costs. Tertiary coil transformers are wired *wye-primary, delta, wye-secondary* and eliminate transmitted third harmonic neutral currents, but cost more due to the larger amounts of material installed.

SECONDARY TRANSFORMERS

Secondary dry-type transformers rated 480 volts delta primary, 208 wye/120-volt, 3-phase, 4-wire secondary (such as those often installed in electrical closets) commonly come furnished with Class H insulation rated for a 150°C rise in temperature. However, if operated continuously at peak load, these transformers could theoretically fail in as little as 2.3 years. Class H, 150°C rise insulated, 3-phase transformers have full load losses of nearly 5% at 30 kVA, 3.5% at 45 kVA and 2.27% for a 300 kVA unit.

Instead, dry-type transformers should be supplied with Class H insulation rated for only an 80°C temperature rise. This change could extend expected life to as much as 25 years at an additional cost of only 10% per transformer which, in addition to reducing life-cyle operating costs, is true *source concept* material and energy resource conservation. Moreover, class H, 80°C transformers are more efficient than the 150°C type due to the larger conductors used to maintain lower operating temperatures; the energy cost savings over 25 years will more than repay the small additional first cost. Class H, 80°C insulated transformers can reduce load losses by as much as 22.5%. Thus, a 30 kVA transformer would lose only 3.69% (of its total kVA rating) and a 45 kVA transformer only 2.74%.

Secondary transformers above 30 kVA should come supplied with six 2½% taps rather than four: two 2½% full capacity above normal (FCAN) and four 2½% full capacity below normal (FCBN). Aluminum windings should neither be considered nor used, especially since common electrical grade aluminum has a much lower tensile strength than copper. The newer triple E aluminum annealed grade does alleviate that problem but is nonetheless prone to more internal I^2R core-and-coil losses than copper. In general practice, transformers below 30 kVA in size are suitable for wall mounting in order to conserve precious floor space, while the additional weight of transformers in sizes 30 kVA and larger are more suited to floor mounting.

MINIMIZING VOLTAGE DROP

Branch circuit wiring should always be designed to minimize the length of conductors (wires) and conduit runs. This not only saves material and reduces labor cost but reduces voltage drop by decreasing the total impedance (resistance) to current flow. Further, current-consuming equipment such as lighting fixtures should not be wired in a series loop where this will increase the distance that current must flow to each fixture. Rather, branch or tap-off connections should be made at the point of power supply to each group of fixtures. Reducing the physical distance that current must flow to each individual fixture in turn reduces the accumulated ampere-feet. The greater the ampere-feet, the greater the voltage drop and I^2R heat losses due to increased energy consumption.

This increased heat loss will increase energy consumption marginally, shorten the life of the electrical equipment, and marginally increase the air conditioning load. These factors will tend to increase life-cycle operating costs for the electrical installation. (See Figures 1-2(A) and 1-2(B).)

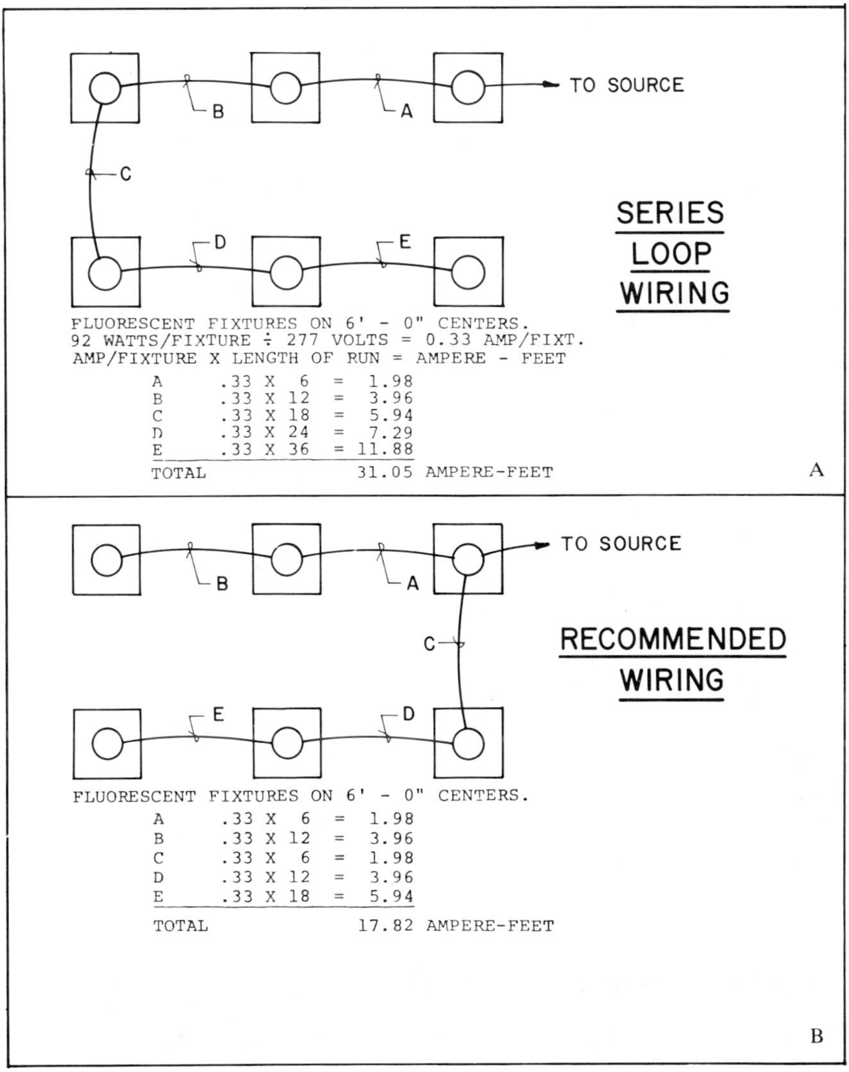

Figure 1-2. Wiring techniques affecting energy usage analysis: (A) series loop wiring, (B) recommended wiring.

RECEPTACLE CIRCUITRY

Specifying standard 4″ × 4″ × 1½″ boxes with a plaster ring for all devices is the most practical solution for commercial and institutional work. The large capacity of the 4-inch square by 1½-inch deep box accommodates large numbers of wires and splices, while the plaster ring permits device mounting. Thus, 3-phase wires (3 circuits) sharing a neutral can be fed through the box while still accommodating a receptacle. Specification of the standard box for all wiring purposes also reduces the electrical contractor's inventory. This also reduces the total energy expenditure for finished goods processed from raw materials.

It also reduces the amount of tool and equipment energy needed to construct the initial installation as well as reducing life-cycle energy use due to the reduction of materials and equipment used and less I^2R energy loss on branch circuit conductors.

Comparison:	(Single Ckts) three 2-wire circuits	=	6 wires
	(3 Ckt groupings) one 4-wire circuit	=	4 wires
	Wires saved	=	2 wires
	Homerun wiring savings	=	33⅓%

Comparison:	(Single Ckts) three ¾" conduits	=	3-¾"C
	(3 Ckt grouping) one ¾" conduit	=	1-¾"C
	Homerun conduit savings	=	66.7%

Electrically, grouping homeruns into threes through the use of 4-inch long by 4-inch wide by 1½-inch deep outlet boxes reduces I^2R energy losses by eliminating imbalanced neutral current. (See Figure 1–3.)

BRANCH WIRING CONDUCTORS AND CONDUIT

Considerations should be given to specifying all branch circuit conductors as minimum #10 AWG rather than the more common #12 AWG. This change can save energy by reducing I^2R line losses in the form of heat and also reduces voltage drop.

We will consider 3 circuits grouped together sharing a neutral forming a 3-phase, 4-wire wiring circuitry. The three circuits feed three balanced loads consisting of either fluorescent or high intensity discharge (HID) lamps such as mercury vapor (MV), metal halide

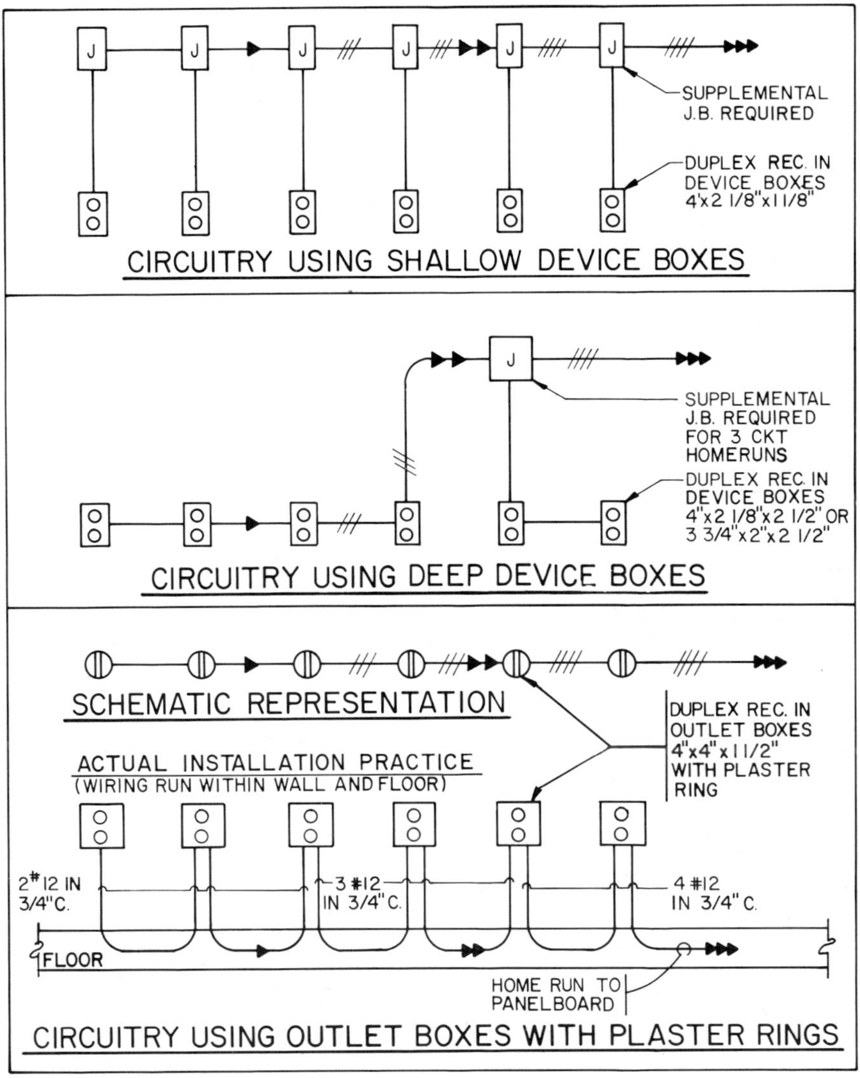

Figure 1–3. Outlet boxes.

(MH), super metal halide (SMH), high pressure sodium (HPS), or high color rendering index HPS (high CRI HPS). As a result of ballast third harmonics the neutral will carry nearly the same current as the phase conductors even though the loads are balanced and there are no out-of-balance neutral currents. We will assume that the average length of each of the conductors is 300 feet. Number 12 AWG type THW copper conductors have a resistance of 0.162 ohm/100 feet so that a fully loaded branch circuit carrying 16 amps would have a voltage drop of 7.8 volts based on V (voltage drop) $=$ I (current on the line) $\times$ R; $V = 16$ amps $\times$ 0.162 ohm/100 feet $\times$ 300 feet. The circuit losses due to copper loss in the form of heat are obtained by the formula $P = I^2R$ or $P =$ (16 amps)^2 ([0.162 ohm/100 feet] $\times$ 300 feet) $\times$ 4 conductors $=$ 497.7 watts. If the general office operating hours are 10 hours per day, 5 days a week and 52 weeks a year, or 2600 hours yearly, this represents 1,293.9 kilowatt hours (kWh). At an energy rate of 6 cents per kWh this comes to $77.64 per year in line losses.

If we consider the same circuit but instead use four #10 conductors, they have a resistance value of 0.1018 ohm/100 feet which will produce a voltage drop of only 4.89 volts. Its line losses for the year tally up to (16 amps)^2 ([0.1018 ohm/100 feet] $\times$ 300 feet) $\times$ 4 conductors $\times$ 2600 hr./yr. $\times$ 6¢/kWh $=$ $48.79 per year.

Four lengths of conductor 300 feet long represent 1200 feet of #12 THW at a cost of $300. The same 1200 feet of #10 THW costs $360. The difference in added material cost is thus $60, but the difference in cost of wasted energy consumed was $28.85 per year. As a result, the payback is just over two years, or 25 months.

Also, all conduit should be sized a minimum of ¾ inch inside diameter (ID) in lieu of the more common practice of using ½ inch ID in order to facilitate the pulling of conductors into these raceways thus reducing labor cost and defraying some of the added cost of materials. This concept can facilitate future changes and expansion and save considerable money over the life cycle of a given facility. Payback is estimated at approximately seven years.

BRANCH DISTRIBUTION IN UNDERFLOOR DUCTS

Telephone wiring and branch circuit wiring from the 120/208-volt power panels installed in the closets that are located on individual floors should be considered for distribution through an underfloor duct system. (See Figure 1–4.) The underfloor ducts can be provided with preset inserts at 60-inch or other modular intervals and will

Figure 1-4. The above overall cutaway isometric graphic illustration clearly shows the interrelationship of the tapway trench header duct which feeds the various services to the cellular floor underfloor ducts. The cellular duct cells in turn feed their various services, i.e., electrical branch wiring to receptacles, multipair, multiconductor telephone lines to telephone amphenol outlets and similar telephone service lines used to feed computer relay terminals (CRTs) and video display terminals (VDTs). These receptacles, telephone amphenols, CRT and VDT terminals are located within preset multiservice flush outlet boxes. These preset inserts are factory located and are locked into the prepunched cellular deck (based upon approved project drawing) prior to the finishing concrete pour. If only the underside of the cellular deck is enclosed, it then can serve as an active cell for any one of the various services. If the top of the cellular deck's flutes, as well as the underside of the cells, is enclosed, it can then serve to bring the various services out of electrical and telephone closets, etc., and feed them vertically and directly into the trench header duct system. The trench header, as pictured here, has a central compartment for the distribution of the electrical convenience receptacle circuitry. The two outside trench feeders can be used for supplying either telephone service and/or CRTs and VDTs. (COURTESY OF H. H. ROBERTSON COMPANY.)

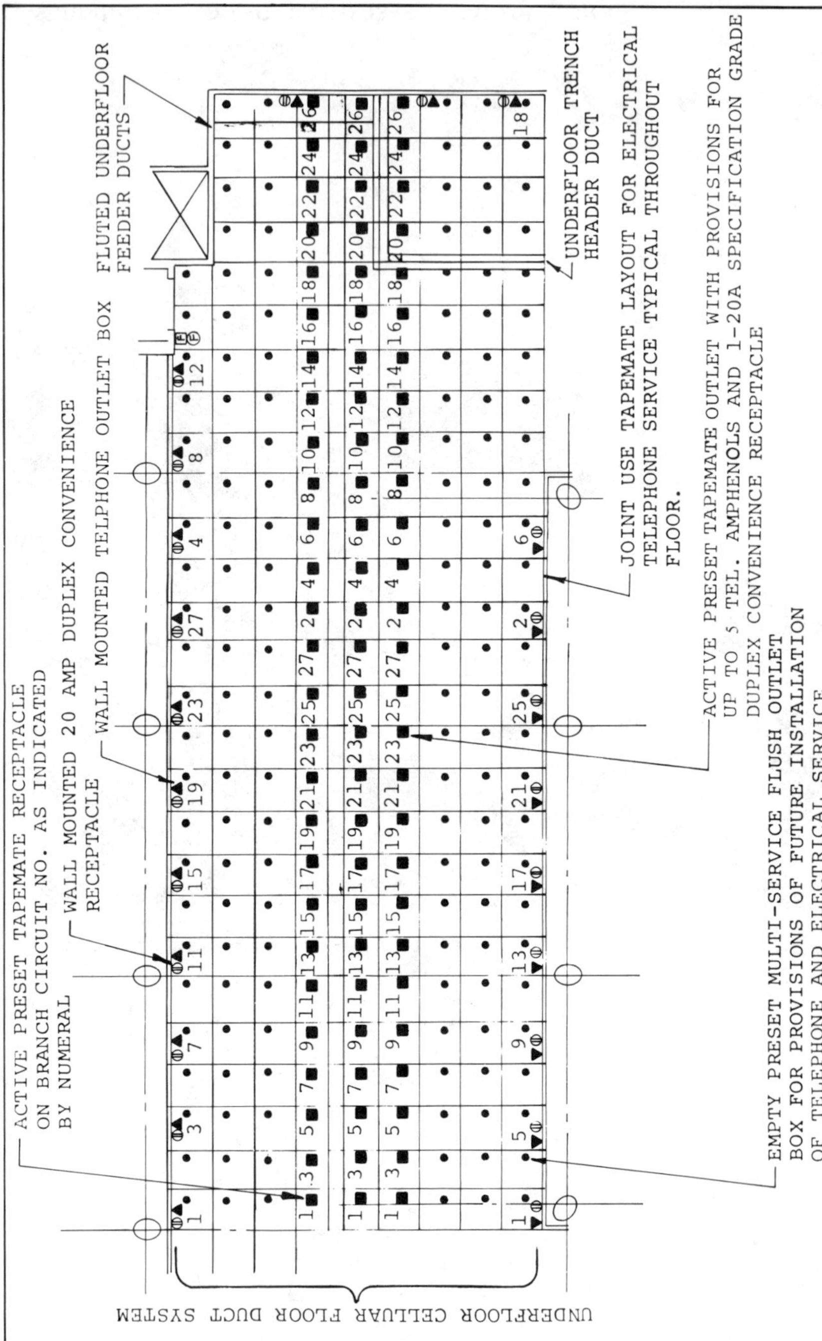

Figure 1–5(A). Underfloor duct system.

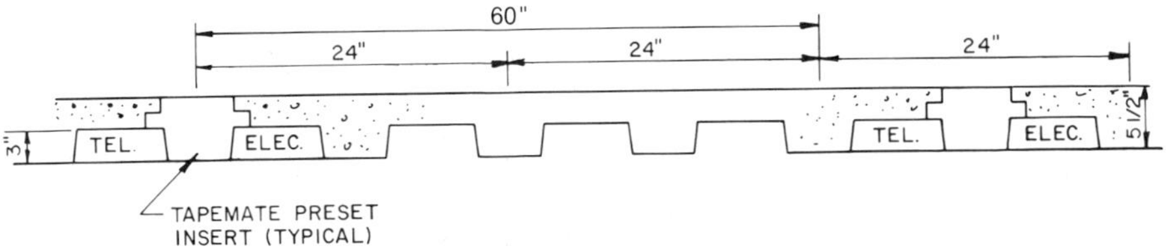

Figure 1-5(B). Section through cellular floor.

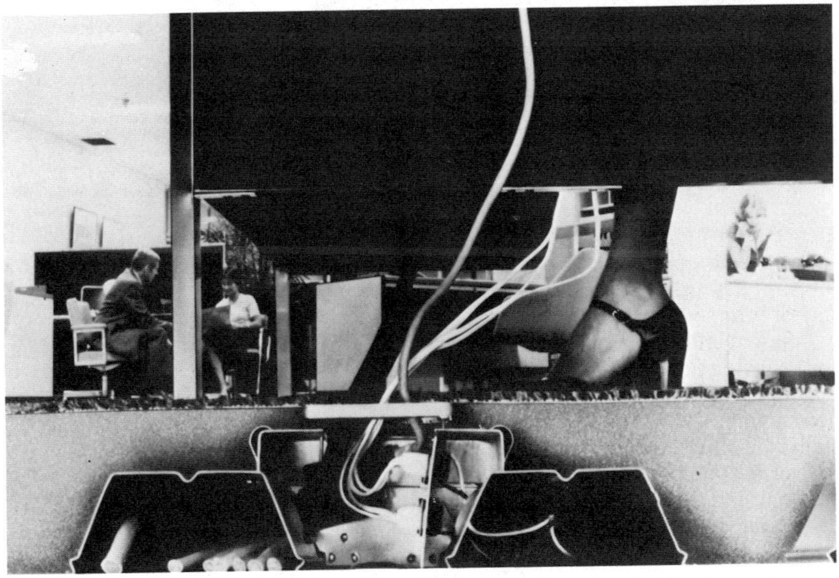

Figure 1-6(A). The cutaway model above shows the interfacing of the activated cells of a cellular floor to its respective activated preset multiservice flush outlet box. This particular activated preset insert is utilized for duplex and/or quadruplex convenience receptacles and termination of several multipair, 25 pair, communication cables into telephone amphenols. Note the isolation barrier in the activated preset insert used to mount the receptacles and isolate the electrical service from the telephone compartment area of the box. This isolation barrier and support thus complies with the National Electrical Code by segregating these two different services. (COURTESY OF H. H. ROBERTSON COMPANY.)

Figure 1-6(B). The H. H. Robertson Q-floor/Taproute system with the Tapmate II (K.E.B. series) preset multi-service flush outlet box is supplied by the H. H. Robertson Company as part of their system, and is normally installed by an electrical contractor sometimes under subcontract to the structural contractor. Those units that are not needed initially or which are not dedicated for immediate use upon occupancy of the facility are normally left deactivated in that no duplex receptacles, telephone amphenols or communication connections are installed in them. An activated unit can accommodate up to two duplex convenience receptacle outlets which are normally installed by the electrical contractor. In addition, an activated unit can handle up to five 25-pair telephone cable amphenol terminal connections which are generally installed by the telephone company. The system can accommodate a public automatic branch exchange system, a Centrex or a less complicated and sophisticated Horizon telephone system. If a private telephone system is installed, ordinarily the manufacturer of that system will be the installer of the multipair telephone cables and desk sets unless the electrical contractor is allowed and elects to install it himself. (COURTESY OF H. H. ROBERTSON COMPANY.)

accept system outlets such as duplex receptacles and telephone amphenols. (See Figures 1-5(A) and 1-5(B).) Thus, power and telephone service can be available at regular intervals, which will offer the utmost flexibility for space planning and work station location.

Figure 1-6(C). After the preset insert multi-service flush outlet box has been activated and its services extended to the work station, then a carpeted enclosure plate can be snapped shut in-place, leaving a smooth, basically uninterrupted carpet area that is both aesthetically pleasing as well as practical and convenient. In the Greater Washington Metropolitan Area the cost of activation of a preset insert box is $40 per unit. The national average is around $30 per activated unit. (COURTESY OF H. H. ROBERTSON COMPANY.)

(See Figure 1-6.) Conductors used for receptacles and appliance branch circuits thus should be #10 AWG in lieu of the normal #12 AWG, in order to comply with the National Electrical Code, Note 8 to Tables 310-16 through 310-19. This note requires that when 25 to 42 conductors are run in a single raceway (cell compartment), the ampere loading shall be only 60% of the maximum permitted.

ITEM	MAXIMUM AMPACITY	60% AMPACITY
#12 wire	20	12
#10 wire	30	18
20-amp circuit breaker	16 (Continuous rating)	--

Therefore, if #10 conductors are used, their deration to loading capacity will exceed the continuous 16-ampere rating of the 20-ampere branch circuit breaker, while still maintaining a margin of capacity useful to counteract voltage drop and I^2R energy losses. Conductors for the telephone system should be sized as required by the telephone system supplier.

Several types of cellular underfloor duct systems to consider are the H. H. Robertson cellular metal floor, Enryco, Roll Form and Stresscon Industries cellular floor systems used in conjunction with structural steel construction for floors above grade, or concrete for slab-on-grade construction. The Enryco Type N-R-G system is by Inland Ryerson, and Deck Systems is a division of Roll Form Products. The floor slabs will retain their NFPA fire rating if the floor is not penetrated more than once every 7 square feet. (The maximum number of penetrations permitted by UL is one per 7 square feet.)

The location and access to the trench header duct should be designed in such a way that the branch circuit systems will be without splice connections in the trench header, thereby negating any NEC access requirements as well as reducing normal access requirements that might be required very infrequently if at all (perhaps once every four years). The trench header cover can be installed in 24-inch by 36-inch sections so that access at any point is convenient, and partitions can be installed across (but not along) the trench header without affecting accessibility. If located in the central corridors, the trench header can be ideally suited to serve up to 10 receptacles per circuit, the maximum number permitted by the National Electrical Code. This fact, combined with wiring the entire length of each cell to permit rapid activation of outlets in the future, will offer maximum flexibility. (See Figures 1–5(A) and 1–5(B).)

The importance of prewiring should be stressed in the original design in order to maintain low life-cycle costs for the system. With prewiring, all wiring can be in place and in the trench header as well as the cells before occupancy of the building by the owner, and new electrical and telephone outlets can be installed quickly and cheaply (in less than an hour for $30–$40) at any time. Without prewiring, activation of new outlets would in many cases require the installation of new wiring from the service closet to the outlet location, necessitating access to the trench header and disrupting office routine all at a much higher cost, and energy wastage in scrapped materials.

PANELBOARD BRANCH FEEDS IN RACEWAYS

Some electrical contractors routinely run all branch circuit conductors from a panelboard through one or two large common raceways to a wiring trough above a hung ceiling where the individual branch circuit conduits then diverge. (See Figures 1-7 and 1-8.) This practice is usually justified on aesthetic or practical grounds as making a neater or more compact installation.

The 1981 National Electrical Code Article 384–15 limits the number of overcurrent devices in one panel to 42 poles, excluding the main overcurrent device. Assuming 42 single-phase branch circuits, and

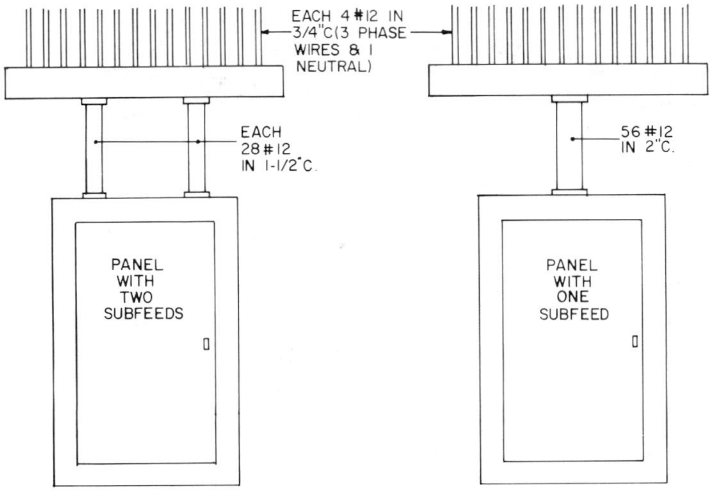

ASSUMING A 42 POLE PANEL, THE MINIMUM NUMBER OF CONDUCTORS TO THE WIRE TROUGH WOULD BE 56 (FOR CIRCUITS GROUPED IN THREES WITH A COMMON NEUTRAL: 42 PHASE WIRES PLUS 14 NEUTRALS). THE MAXIMUM NUMBER WOULD BE 84 (42 SEPARATE CIRCUITS: 42 PHASE WIRES PLUS 42 NEUTRALS).

Figure 1-7. Branch circuits.

Figure 1–8. Actual installation utilizing the installation concept of one or two subfed raceways from the panelboard to an overhead wireway or wiring trough.

grouping circuits in threes with a common neutral, there may be as many as 56 branch circuit conductors—42 phase conductors plus 14 neutrals. If for some reason 42 separate branch feeds are made, then there could be 84 branch conductors—42 phase wires and 42 neutrals.

If a fault occurs to a phase wire in a single group of 3-phase wires and a neutral (4 #12, $\frac{3}{4}$"C) then only $\frac{3}{42}$ (7.1%) of the panel circuits are affected at one time. However, if a fault occurs in the subfeed of a panel having two conduit subfeeds, half of the panel circuits are affected. With a single subfeed conduit, a fault in a single-phase wire could cause the burnout of the remaining 41 branch circuits. The second problem, noncompliance with the NEC, arises because most engineers and electrical contractors do not follow derating requirements. Note 8 to Tables 310–16 through 310–19, "more than three conductors in a raceway or cable," specifies the following:

NUMBER OF CONDUCTORS	PERCENT OF VALUES IN EITHER 310–16 AND 310–18
4–6	80
7–24	70
25–42	60
43 and above	50

The neutral of a 3-phase, 4-wire circuit supplying electric discharge lighting is also considered as a current-carrying conductor, by note 10(c). The neutral of a 3-phase, 4-wire circuit supplying receptacles or other noninductive or noncapacitive miscellaneous power is not considered a current-carrying conductor.

As a result, an overhead fluorescent lighting system supplied by four #12 conductors in $\frac{3}{4}$″C should be derated to 80%. The twin subfeed panel's conductors should be derated to 60%, and the single subfeed panel's conductors to 50%. The alternative is to oversize the conductors, which is also a poor utilization of resources and energy.

POWER FACTOR CORRECTING

All electrical equipment and devices used should be specified with the highest power (PF) available. In particular, all fluorescent and HID lighting fixture ballasts should have a rated PF greater than 90%.

Where individual pieces of equipment consuming 1500 watts of energy or more have a rated PF below 90%, capacitors should be installed to correct the PF to 95% or better. All motors of 2 hp (approximately 2715 VA) and above should come supplied with PF corrective devices.

Power-factor correction should be employed to conserve energy and reduce life-cycle costs. It increases building distribution capacity (more kW from the same transformers), improves voltage regulation, and reduces power losses. (See Figure 1–9.)

Motors and other inductive equipment require two types of electrical energy. One type is working or power-producing current,

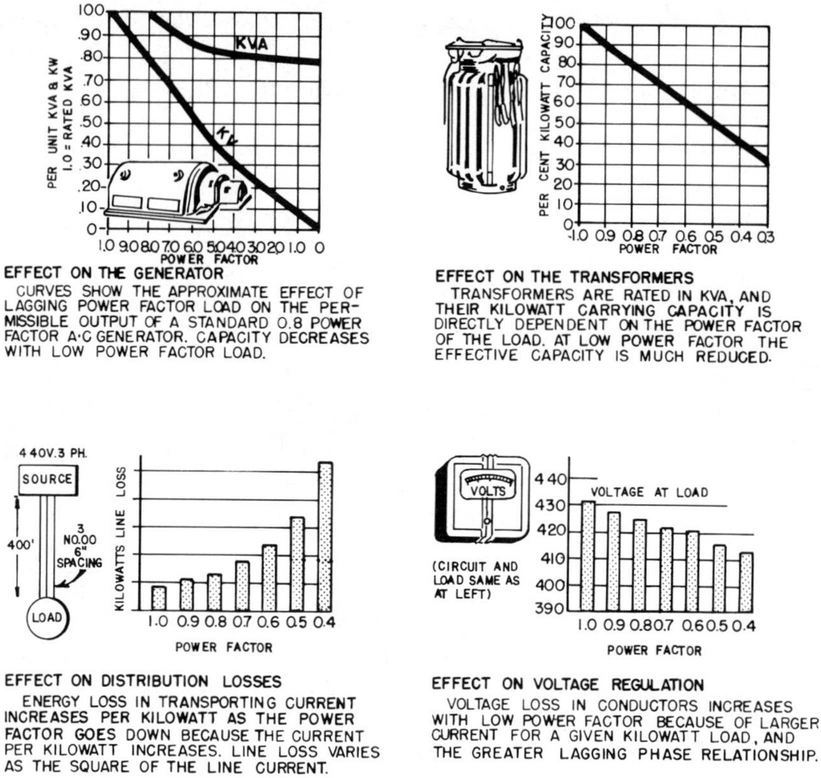

Figure 1-9. Effect of lagging power factor an AC equipment.

measured in kilowatts (kW). The second type is magnetizing current, measured in kilovoltamps-reactive (kVAR), which produces no useful work but is nonetheless necessary to the functioning of the equipment. Capacitors can be installed at motors and other inductive equipment to supply the necessary kilovars of magnetizing current directly at the loads rather than throughout the distribution system. Thus, PF correction can be easily retrofitted to existing buildings.

Low power-factor equipment consumes more kVAR than those rated at high power factors (0.8 or greater). Low power factor is also detrimental to the building electrical system in the following ways:

1. It cuts down system loadability, that is, it reduces the capacity of the power system to carry kilowatts (kW = kVA × PF). The

capacity of all apparatus is determined by the kVA it can carry; hence, larger transformers, feeders and switches must be provided for each kilowatt of load when the power factor is low.

2. Low power factor means more current per kilowatt; so each kilowatt of power carries a higher burden of I^2R line losses in the form of heat.

3. Low power factor may depress the voltage, with a detrimental reduction in the output of practically all electrical apparatus and more energy wastage.

Although capacitors can be used at the service entrance to correct PF, this has no effect on the overall system's poor power factor. Thus, capacitors are most effective when installed directly at points of load, as previously described. A simple illustration of the effect of capacitors is a motor drawing 10 amps of current. At 80% power factor, only 8 amperes produce working horsepower, while 2 amperes serve as magnetizing current. (See Figure 1-10(A).)

In Figure 1-10(B) a capacitor has been added to the feeder to supply the kilovars directly at the motor. Now the feeder must deliver only 8 amperes for exactly the same horsepower output. In this way, capacitors reduce the line current and make possible the addition of

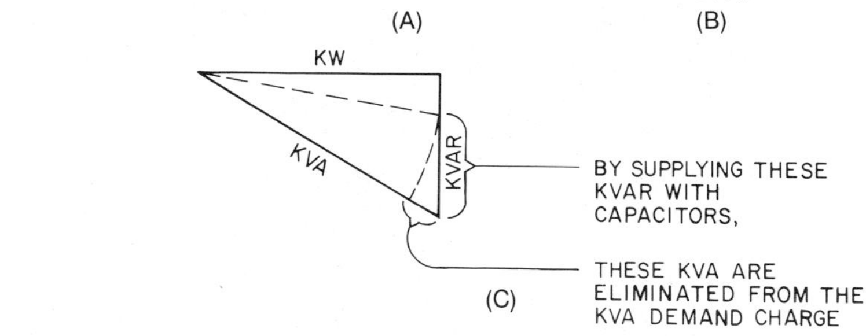

Figure 1-10. Power factor.

more electrical equipment to fully loaded feeders. If this same current fed four motors at 80% power factor, then the use of capacitors would permit another motor to be added to the line with no increase in the total current demand.

Capacitors release system capacity so that more motors, lighting and other loads may be added to the system without overloading transformers and other distribution equipment. (See Figure 1-10(C).) This system release continues as you add capacitors up to the point at which 100% power factor is reached. However, more kVAR of capacitors are required per kW of released capacity as the power factor increases. Capacitors should, therefore, be added until the cost of capacitors to release each kW of capacity equals the cost per kW of new transformers and distribution equipment. Raising the power factor to at least 95% and up to 98% is usually economical.

Capacitors to correct PF cost from \$14 to \$38 per kVAR for a 277/480-volt system and \$29 to \$56 per kVAR at 120/208 volts in going from 25 kVAR (least costly in dollars per kVAR) to 5 kVAR (most costly per kVAR). This is another sound reason for the use of higher voltage distribution systems.

By improving the power factor of a squirrel cage induction motor from 0.8 to 0.95, substantial energy savings can be realized, as the following analysis indicates.

Voltage = current × resistance ($V = IR$) and by algebraic manipulation, $I = V/R$ and $R = V/I$. Real power (P) is the *time rate of energy consumption* and equals voltage times current when both quantities are in phase or have unity power factor (PF = 1). Thus, power in watts equals volts times amps times cos θ.

$$P = V(I) \text{ @ } PF = 1$$

$$\text{power} = VI = (IR)I = I^2R$$

$$\text{current } (I) = \frac{\text{voltamps,}}{\text{voltage}} \text{ or, } I = \frac{VA}{V} \quad \text{where } A = \text{Current in amperes}$$

$$\text{voltamps } (VA) = \frac{\text{watts,}}{PF} \text{ or } VA = \frac{\text{W}}{PF}$$

and

$$I = \frac{VA}{V} = \frac{\text{W}}{V(PF)} = \frac{\text{watts}}{V(PF)}$$

thus

$$I(0.8\,\text{PF}) = \frac{\text{W}}{V(0.8)} = \frac{1.25\,\text{W},}{V} \quad \text{or, } I(0.8\text{PF}) = 1.25I \text{ at unity power factor.}$$

The current at 0.8 PF has been increased by 1.25 times the current at unity power factor ($PF = 1$).

Thus, the power @ (0.8 PF) $= \left(\frac{I}{.8}\right)^2 R = (1.25I)^2 R = 1.5625I^2R$

and, power @ 0.95 PF $= \left(\frac{I}{.95}\right)^2 R = (1.0526I)^2 R = 1.108I^2R$

The difference in power consumed at 0.95 PF versus 0.8 PF equals

$$\frac{(1.108 - 1.5625)}{1.5625} \times 100 = -29.088\%, \text{ or } 29.1\% \text{ less energy used}$$

or, mathematically stated, the algebraic letters I and R are constant in value and only their coefficients change. Thus, the difference in percentage is

$$\left[\frac{\left(\frac{1}{0.95}\right)^2 - \left(\frac{1}{0.8}\right)^2}{\left(\frac{1}{0.8}\right)^2}\right] \times 100 = -29.1\%$$

A variation on the previous formula is that $W = VI$ at unity power factor ($PF = 1$) and $P = I^2R$, thus

$$I = \frac{W}{V(PF)}$$

$$P = I^2R \text{ so that } P = \left[\left(\frac{W}{V(PF)}\right)^2 - \left(\frac{W}{V(PF)}\right)^2\right] R$$

or the difference in energy consumed would be $P = [(I/0.8)^2 - (I/0.95)^2]R$, which concurs with our previous analysis.

Another useful formula is the following: Percentage reduction of power losses equals $100 - [100(0.8/0.95)^2 = 100 - [100(0.84211)^2 = 100 - [100(.70914)] = 100 - 70.914 = 29.1\%$, which correlates with the preceding solution derived from another approach.

By reducing the current utilized by electric motors (inductive devices) through improving their power factor, that current that is in phase with the voltage ($P = VI \cos \theta$) represents the real power (*time rate of energy consumed*) that can be reduced and conserved. Capacitors reduce system losses by reducing total current and power flowing through that system. For example, a 20% reduction of the total current will cut the power (energy) losses 36%. Savings from power loss recovery alone can give you an annual return on a capacitor investment of as much as 20% or more. In some facilities it can be paid back in a matter of years.

We will make an analysis of energy saving for a 20 hp motor utilizing power factor corrective equipment in order to bring its industry average power factor (PF) of 80% (0.8) up to 95% (0.95). By improving its PF from 0.8 to 0.95 we wish to see what energy savings may be accrued both initially and in its life-cycle cost savings as well. The prevailing average power charge is about 6¢ per kilowatt hour for small commercial installations having a fairly low kVA demand.

One horsepower equals 746 watts; thus a 20-hp motor will utilize 20 hp × 746 watts/hp = 14,920 watts of electrical energy.

With an improvement in PF from 0.8 to 0.95 and 0.68 efficiency we will reduce the current by the following amount: for a 120/208-volt, 3-phase, 4-wire, wye-connected system we know that the phase-to-phase (line-to-line) voltage will be 120 volts times the square root of 3 equals 208 volts, and that the system's current is derived by dividing the total system wattage by 208 V_{LL} (line-to-line) times the square root of 3. Thus,

$$\frac{14{,}920 \text{ watts}}{0.8 \text{ PF} \times 0.68 \text{ EFF.}} = 27{,}426.47 \text{ VA} \qquad \text{EFF.} = \text{Efficiency}$$

$$\frac{27{,}426.47 \text{ VA}}{208 \text{ V}_{LL} \sqrt{3}} = \frac{27{,}426.47 \text{ VA}}{360 \text{ V}} = 76.184 \text{ amps}$$

$$\frac{14{,}920 \text{ watts}}{0.95 \text{ PF} \times 0.68 \text{ EFF.}} = 23{,}095.975 \text{ VA}$$

$$\frac{23{,}095.975 \text{ VA}}{208 \text{ V}_{\text{LL}} \sqrt{3}} = \frac{23{,}095.975 \text{ VA}}{360 \text{ V}} = 64.155 \text{ amps}$$

Therefore, the difference in current is 76.184 amps − 64.155 amps = 12.058 amps. (See Figures 1–11 and 1–12.)

The vector in the denominator is 360 volts and represents the three 120-volt phase to neutral vectors of the system, i.e., 3(120 volts) = 360 volts. From a mathematical point of view, 208 volts line-to-line was arrived at by multiplying 120 volts times the square root of 3; thus 208 V_{LL} times the square root of 3 is:

$$\begin{aligned} 120 \text{ V} \sqrt{3}\,\sqrt{3} &= 120 \text{ V}(\sqrt{3})^2 = 120 \text{ V} \times 3^{2/2} = 120 \text{ V} \times 3^1 \\ &= 120 \text{ V}(3) = 360 \text{ volts (See Figure 1–11.)} \end{aligned}$$

There are spin-off savings such as increased life due to the reduction of I^2R energy losses in the line in the form of added heat which decreases the life of all electrical equipment. We will consider that the branch feeder consists of 100 feet of #4 AWG conductors having 0.0321 ohm of resistance per conductor. Since these are balanced 3-phase inductive loads (squirrel cage AC motors) connected in delta or possibly in wye without a neutral connection, the current will have a one-way run. Even if a neutral was present and connected, its current would be a vector summation of zero barring any third harmonic field induced current. The current was reduced by 12.058 amps; therefore, we will have a one-way run for the conductor (one per phase) with a 100-foot length from the point of origin (panelboard) to the point of load termination (motor). For the same hours of operation as the motor it serves, we will have 3120 hours of use per year. The power saved will be $P = I^2R$ for 3 conductors:

$$[(76.184 \text{ amps})^2 - (64.155 \text{ amps})^2] \times 3 \times 0.0321 \text{ ohm}$$

$$= (5{,}804.0019 \text{ amps}^2 - 4{,}115.684 \text{ amps}^2) \times 0.0963 \text{ ohm}$$

$$= (1{,}688.1378 \text{ amps}^2 \times 0.0963 \text{ ohm}) = 162.567 \text{ watts.}$$

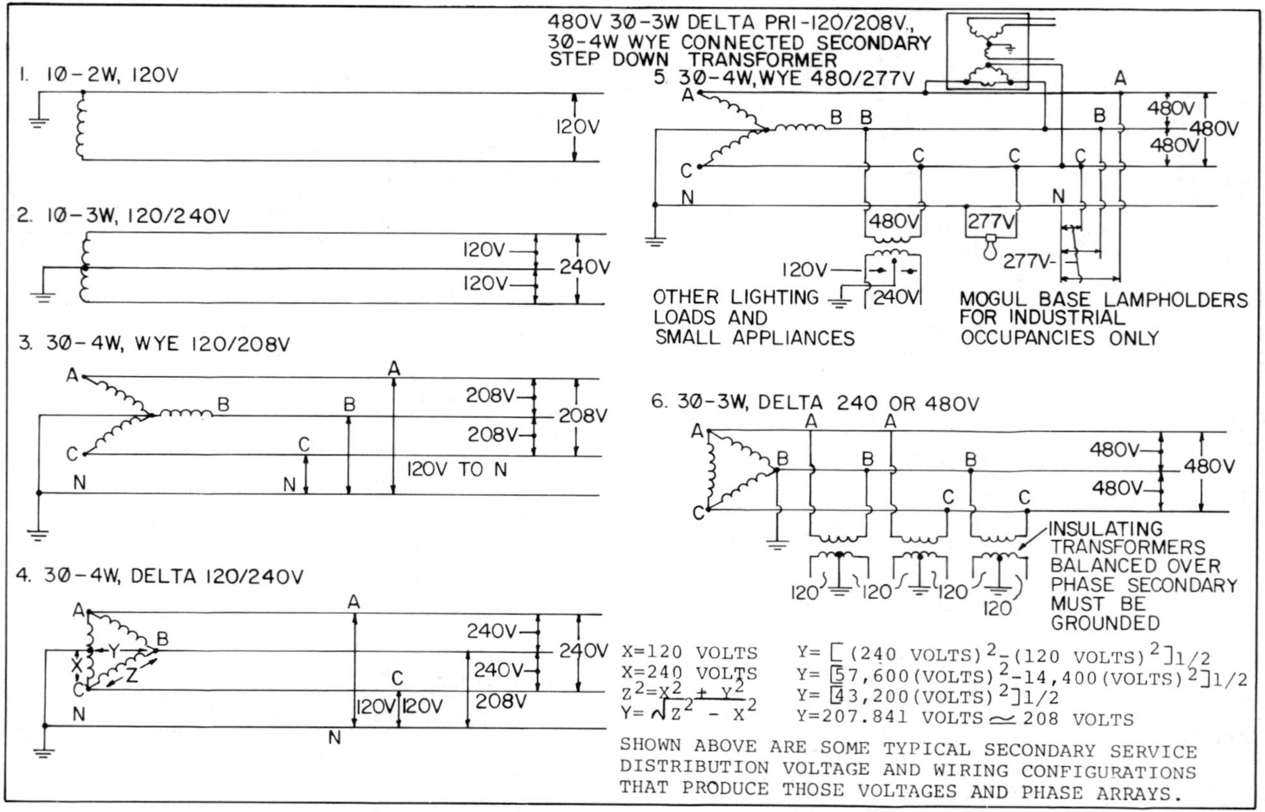

Figure 1–11. Transformer and voltage configurations.

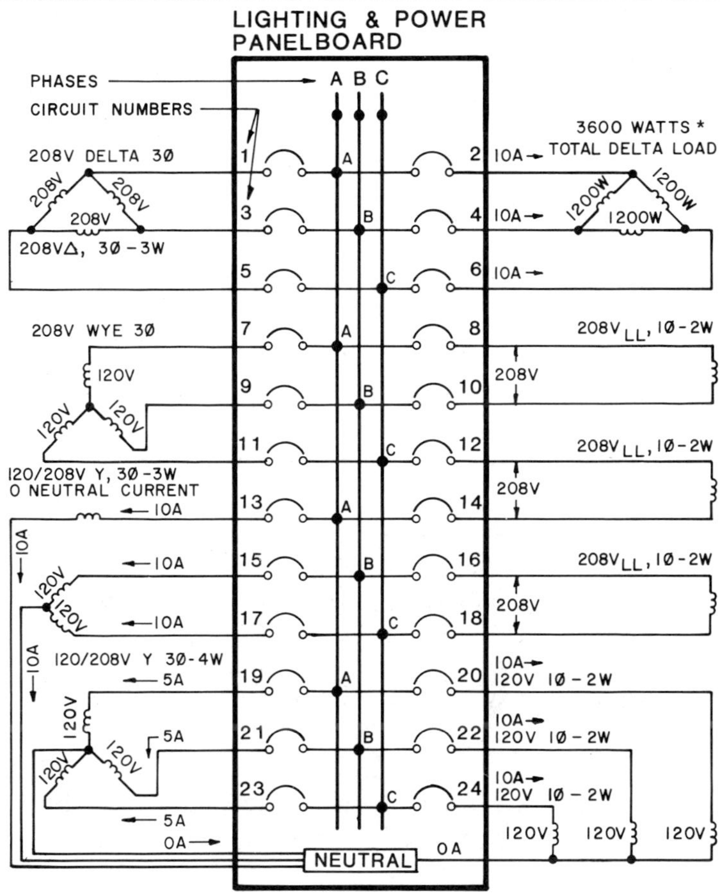

*
JUST AS 10 AMPS TIMES 120 VOLTS (AT UNITY POWER FACTOR) EQUALS 1,200 WATTS SO TOO DOES 1,200 WATTS AT 120 VOLTS LINE TO NEUTRAL (PF=1) EQUALS 10 AMPS OF LINE CURRENT. BY THE SAME TOKEN A BALANCED 3 PHASE DELTA CONNECTED LOAD OF 3,600 WATTS (PF=1) AT 208 VOLTS (LINE TO LINE) 3 PHASE, 3 WIRE EQUALS A LINE CURRENT OF 10 AMPS PER EACH OF 3 PHASES. 208 VOLTS (LINE TO LINE) EQUALS TWO 120 VOLT LINE VOLTAGES ADDED VECTORALLY TO PRODUCE $120V_{LL}\sqrt{3}$ OR $208V_{LL}$ THEREFORE THE LINE CURRENT OF A 3,600 WATT 3Ø-3W DELTA CONNECTED LOAD $= \frac{3{,}600 \text{ WATTS}}{208V_{LL}\sqrt{3}} = \frac{3{,}600 \text{ WATTS}}{360 \text{ VOLTS}} = 10$ AMPS

Figure 1-12. Voltage and phase configurations.

If the motor runs 12 hours a day, 5 days a week for 52 weeks a year, then it will operate for 3120 hours in a year.

$$\frac{162.567 \text{ watts} \times 3120 \text{ hr/yr}}{1000 \text{ watts/kW}} \times \$.06/\text{kWh} = 507.21 \text{ kWh} \times \$.06/\text{kWh} =$$

\$30.43 of energy saved.

For this analysis, the distribution panel from which the motor is fed has been chosen as being 200 feet away from its source of power. The distribution panel is in turn fed by four 500 MCM feeder conductors of which only a one-way run with 3 of the 500 MCM conductors is applicable as per previous analysis. Two-hundred feet of 500 MCM conductors has 0.00588 ohm of resistance; so that energy savings are:

$$[(76.148 \text{ amps})^2 - (64.155 \text{ amps})^2] \times 3 \text{ conductors} \times 0.00588 \text{ ohm} = (5{,}804.0019 \text{ amps}^2 - 4{,}115.864 \text{ amps}^2) \times 0.01764 \text{ ohm} = (1{,}688.1378 \text{ amps}^2) \times 0.01764 \text{ ohm} = 29.778 \text{ watts}$$

$$\frac{29.778 \text{ watts} \times 3120 \text{ hrs./yr.}}{1000 \text{ watts/kW}} \times \$0.06/\text{kWh} = 92.9097 \text{ kWh} \times \$0.06/\text{kWh} = \$5.57.$$

We thus have a total I^2R energy savings attributable to the branch and feeder conductors of \$30.43 + \$5.57 = \$36.00. The additional first cost of adding *PF* corrective capacitors can be paid back within 10 years, depending on the manufacturer selected and their cost to the owner.

A first-cost charge for 7.5 kVAR of 240-volt power factor corrective capacitors will range from \$202 for Ronk Electrical Industries, Inc. supplied equipment to \$380 as supplied by General Electric. An electrical installation and connection charge of \$40 can be expected. Thus, the installed first cost will be anywhere from \$242 to \$420, depending on manufacturer. Capacitive power factor corrective devices can be expected to last from 12 to 15 years.

With an expected life of up to 15 years for *PF* corrective devices, large accrued savings can be expected. Based on the first cost of equipment and installation of \$242 to \$420, your return on investment will be from 8.57%, [(\$36 total energy cost savings × 100%)/\$420 installation cost] to as high as 14.87%, [($36 total energy cost savings × 100%)/$242 installation cost].

However, under normal operating conditions for various sized motors ranging from 2 to 50 hp, Capacitor Monitor Company states that most in-plant capacitors earn 50% to 300% return-on-investment.

There may still be some confusion or concern as to how we could possibly save this much energy by just correcting the PF. We will thus extend this analysis one step further in order to clarify how this is all possible.

When considering direct current (DC) systems, there are primarily three variables: current (I), voltage (V) and resistance (R); time (t) is a fourth variable but does not apply to steady-state conditions. Their relationship to each other can be expressed mathematically as voltage equals current times resistance ($V = IR$). The power delivered to a circuit or electrical system is equal to the product of the current through the circuit and the voltage drop across the circuit; or their relationship can be extended to cover varying currents and voltages simultaneously. Thus, power equals the product of voltage times current ($P = VI$).

Thus, if $V = IR$ as previously stated, then an algebraic rearrangement allows that $I = V/R$ and $R = V/I$. By having solved for one variable we can mathematically transpose this new identity and obtain $P = (IR)I = I^2R$ or $P = V(V/R) = V^2/R$, each of which is equally as correct as $P = VI$.

In the case of an alternating current (AC) system, we will deal with a steady-state condition which is based upon the use of a forced response which is repetitive or periodic. Thus, a sinusoidal forcing function applied to a network equation is a steady-state response.

Basically, all our domestic power supplies are at frequencies of 60 Hertz or, as it is sometimes referred to, 60 cycles per second. These power supplies are sinusoidal forcing functions.

In AC systems there are several additional and different variables; the voltage (V) remains the same, but the total resistance to current is a much larger entity comprised of several variables. The total AC resistance is referred to as the impedance (Z) and is comprised of pure resistance (R) and reactance (X), while the reactance is comprised of inductive reactance (X_L) and capacitive reactance (X_C). The impedance from an inductive device is expressed as $Z_L = JX_L$ where

$X_L = WL$, whereby J = $\sqrt{-1}$ (the square root of minus one which is an imaginary number) and W or radian velocity is expressed as 2 times pi (π) times the frequency (F) (60 Hz), or $2\pi FL$, and likewise the impedance of a capacitor is expressed as $Z_C = 1/JWC$, and if multiplied and divided simultaneously by the mathematical entity J (which does not change the equation), then

$$Z_C = \frac{1}{JWC} \times \frac{J}{J} = \frac{J}{J^2WC} = \frac{J}{\sqrt{-1}\ \sqrt{-1}\ WC} = \frac{J}{(-1)^{2/2}WC}$$

$$= \frac{J}{-WC}$$

since $J = \sqrt{-1}$.

Thus the total system impedance Z equals $\left[R^2 + \left(WL - \dfrac{1}{WC}\right)^2\right]^{1/2}$

or $Z = \left[R^2 + \left(WL - 1/WC\right)^2\right]^{1/2}$ and current I equals $\dfrac{V}{Z}$

$$= \frac{V}{[R^2 + (WL - 1/WC)^2]^{1/2}}$$

The real power is, of course, that current which is in phase with the voltage and that portion that is in phase is represented by the value of the power factor. Thus a 5 kVA load having a current-voltage phase difference of 36.9° represents a PF of 0.8 ($PF = \cos 36.9°$). This is shown graphically as

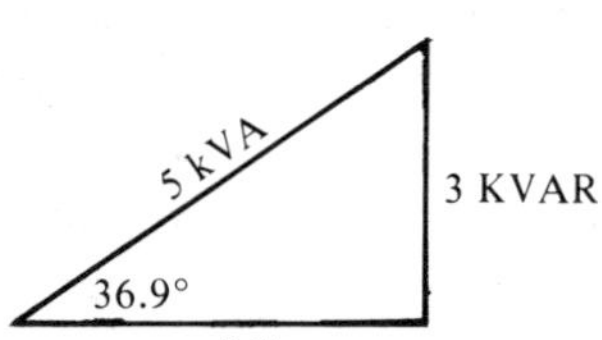

and represents 4kW of real load, 3 kilovoltamp-reactive units (kVAR) for a total of 5 kVA.

Therefore, there are only 4 kW of useful load out of a total of 5 kVA. Armed with this knowledge we know that the current is out of phase with the voltage by 36.9° and therefore, whether the current is leading or lagging, only 0.8 of the current is in phase with the voltage. Thus, 120 volts times 10 amps equals 1200 voltamps; but 120 volts times 10 amps @ 0.8 PF equals only 8 amps in phase with the voltage to produce only 960 watts of real power. Since motors are made up of inductance expressed as JWL the current is derived by $I = V/Z$ where Z equals $(R^2 + J^2)^{1/2}$, the $V/\text{JWL} = -JI$, or $I \angle -90°$. I is lagging by 90° behind the voltage where the load is purely inductive since $-J = -90°$. Since it is not purely inductive but is an impedance made up of some resistance, then the current merely lags behind the voltage at an angle less than 90° but greater than 0°, expressed mathematically as the tangent of J/R where J in this case equals

$$\left(WL - \frac{1}{WC}\right) = \left(\frac{W^2LC - 1}{WC}\right)$$

Let us evaluate a 10 hp motor having an efficiency of 80% (0.8) and an initial industry standard power factor of 80% (0.8) that will be increased to 0.95 PF with the aid of capacitors.

$$\frac{10\text{ hp} \times 746\text{ watts/hp}}{(0.8\text{ PF})\,(0.8\text{ EFF.})} = \frac{7{,}460\text{ watts}}{0.64} = 11{,}656.25\text{ voltamps}$$

$$\frac{10\text{ hp} \times 746\text{ watts/hp}}{(0.95\text{ PF})\,(0.8\text{ EFF.})} = \frac{7{,}460\text{ watts}}{0.76} = 9{,}815.7895\text{ voltamps}$$

$$11{,}656.25\text{ VA} - 9{,}815.7895\text{ VA} = 1{,}840.26\text{ voltamps.}$$

Thus, the reduction in current at 0.95 PF vs. 0.8 PF is

$$\frac{1{,}840.26\text{ VA}}{208\text{ V}_{\text{LL}}\sqrt{3}} = \frac{1{,}840.26\text{ VA}}{360\text{ V}} = 5.1118\text{ amps}$$

or, stated another way, the current to a 10 hp motor at 0.8 PF will be

$$\frac{11656.25\text{ VA}}{208\text{ V}_{\text{LL}}\sqrt{3}} = \frac{11656.25\text{ VA}}{360\text{ V}} = 32.3784\text{ amps}$$

The current to a 10 hp motor at 0.95 PF will be

$$\frac{9815.7895 \text{ VA}}{208 \text{ V}_{LL}\sqrt{3}} = \frac{9815.7895 \text{ VA}}{360 \text{ V}} = 27.266 \text{ amps}$$

or the reduction of current at 0.95 PF over 0.8 PF is 32.3784 amps − 27.266 amps = 5.1124 amps which confirms our above analysis.

This last example once again reinforces our original premise that the accrued savings are based upon a reduction in the total system's current. This releases system capacity and reduces energy consumption due to line losses in the form of (I^2R) heat losses. This heat is deleterious to all electrical equipment causing reductions in useful life as well as marginally increasing the air conditioning load.

As can be seen by the previous analyses, since the *PF* corrective equipment is placed external to the electric squirrel cage motor, that is, it is connected to the load side of the motor starter, it does not affect the current flowing through the motor. Thus, no energy savings are accrued from within the electric motor. The magnetizing current to the motor has been countered by capacitive kVARs nearly equalizing the kVARs of inductance. Thus, the total system component of current throughout the entire electrical distribution system has been reduced. The system losses have also been reduced and finally we have the accrued value of additional freed-up system capacity.

The majority of electric utilities have a penalty clause in their tariffs that places an extra charge for low power factor. There are, however, no standard methods of applying this tariff nor is there a standard charge. Most utilities charge a stepped rate for each kilowatt hour of energy consumed, i.e., the first 6,000 kWh, no demand billing rates; the next 200 hours of use but less than 200,000 kWh, 4¢/kWh; all additional kWh, 2.1¢/kWh; and so on, based upon whether it is a commercial, industrial or residential customer. There are various classes of service based upon a demand charge, voltage characteristics and number of phases, i.e., first 25 kW of month's billing, no charge; next 75 kW monthly demand, $6.25/kW; and in excess of 100 kW/month, $6.00 kW, etc., as well as an adjustment charge based upon peak and nonpeak monthly billing; 120/240-volt, single-phase, (1Ø), three-wire, (3W); 120/208V, 3Ø, 4W; 265/460 V, 3Ø, 4W, etc. In addition, these utilities may charge extra for any power factor below say 85% and this charge may be a weight factor such that at 80% PF you must pay an additional 6.25%, [(85%–80%)/85%]. The penalty may also be calculated as a ½% additional charge for each percent the PF is below 85%. Thus, an industrial complex running at 70% PF with a monthly kWh bill of $3000 will pay an additional ½% (85–70). This computes out to ½%(15) or 7½% of $3,000 which equals an additional $225. If the customer is found to have either a leading or lagging power factor of less than 85% the demand charges may be multiplied by a factor of 1.111.

Some utilities have a stepped rate of charge for the number of hours of use of kilovoltamp-reactive units used, i.e., so many cents for the first 1000 kVARh used and so many more cents for any additional VARhs used. Still other utilities charge for kVAhs of energy used which is the vector summation or resultant of kWh and kVARh expressed simply in mathematical terms as kVAhs $= \sqrt{(\text{kWhs})^2 + (\text{kVARhs})^2}$. Still another approach used has a charge for reactive power used such that for each kVAR in excess of requirement for 85% power factor will cost an additional $0.15 per kVAR.

As a result, any savings accrued by reducing or eliminating a power factor penalty through the use of capacitive *PF* correction equipment must be evaluated on the basis of the tariffs imposed by the utility company supplying the power. Thus, no generalizations can be made as to what kind, type or dollar savings can be accrued through the improvement of PF due solely on the basis of utility company tariff reduction. PF penalty clause reductions thus must be ascertained on a one by one basis, determining who is the supplier of the electric energy and what tariffs, if any, does this utility charge.

The larger manufacturers of power factor corrective devices are General Electric (low voltage under 600 V), Westinghouse Electric Corporation (low voltage under 600 volts—medium to high voltage), McGraw Edison (medium to high voltage application), Kearney (medium to high voltage application). In addition, the following also manufacture *PF* corrective devices: Aerovox, Inc., New Bedford, MA; Arco Electric Products Corp., Shelbyville, IN; BICC Ltd., Houston, TX; Dayton Electric Mfg. Co., Chicago, IL; Elliott Industries Inc., Bossier City, LA; Essex

Group/Controls Div./Steveco Products, Logansport, IN; Federal Pacific Electric Company, Newark, NJ; Fresco Inc., Minneapolis, MN; W. W. Grainger, Inc., Chicago, IL; Motors and Armatures, Inc., Hicksville, NY; Ohio Transformer Corporation, Louisville, OH; H. K. Porter Co., Inc., Electrical Division, Pittsburgh, PA; Sangamo Energy Management Div., Sangamo Weston, Inc., Atlanta, GA; Shallbetter Industries Inc., Golden Valley, MN; Siemens Corporation, Iselin, NJ; Sprague Electric Co., North Adams, MA; Zucker, Myron, Inc., Bloomfield Hills, MI.

ENERGY-EFFICIENT ELECTRIC MOTORS

Energy-efficient, squirrel-cage induction motors, in addition to having higher efficiencies, generally have higher PF's as well. This last fact can substantially reduce first costs through reduction in the amount of capacitors needed to improve the PF. These motors not only save energy; they save money as well, and their payback is often measured in months rather than years. These *energy-efficient* or *high-efficiency* motors can accrue an annual 25% energy and cost savings. Because of their greater efficiency there is less energy wastage in the motor, which prolongs life, since for every 10°C reduction in temperature there is an approximate doubling of life. The larger windings within the motor reduce the internal resistance as well as additional current flow, thus reducing I^2R heat losses, and marginally reducing air conditioning loads while increasing motor efficiency and life. Winding related losses vary with load current and are heat losses due to resistance to this current passing through the rotor and stator conductors and can account for upwards of 60% of the total motor energy losses. There are losses in the form of heat in eddy currents and hysteresis losses which represent another 20% to 25% loss in efficiency. Hysteresis losses are the major component of this loss and are due to lower magnetic quality steel being used, as well as poor steel grain orientation. Special processing of cold roll steel or especially the use of silicon steel (which is far more superior) will improve the electromagnetic properties of the stator core. Core laminate thickness reduction and increasing the core size (preferably its length and not its width) will reduce its flux current and I^2R heat losses.

Friction and windage losses in a motor normally account for another 10% energy loss. These windage losses can be minimized by fabricating the rotor and stator windings with high conductivity copper of larger cross-sectional area (increased diameter conductors). With lower I^2R losses the motor's fan size and windage loss can be cut down.

High efficiency motors are now perfected and are state-of-the-art. Generally, these motors have an additional premium cost of from 5% to 25%. A functional analysis based on life-cycle costs will show that energy efficient motors have a payback period of from 0.5 year to about 0.71 year, depending on the size of the motor and the degree of higher efficiency. (See Tables 1–1, 1–2 and 1–3) A quick and easy analysis method of evaluating your potential annual energy cost savings for a specific higher efficiency motor of a given hp size is the formula kW (saved) times hours of usage times the cost per kWh, where

$$\text{kW (saved)} = \frac{\text{hp} \times 746 \text{ watts/hp}}{1000 \text{ watts/kW}} \times \left[\frac{1}{A} - \frac{1}{B}\right]$$

and A = efficiency of a standard NEMA motor and
B = efficiency of a higher efficiency motor.

The following example is based upon an analysis of a 7.5 hp motor having a standard NEMA motor efficiency of 84.3% versus a 7.5 hp higher efficiency motor rated 90.2% efficient. A 4000 hour annual operation is assumed and an energy cost of 6¢/kWH is used.

$$\text{Savings} = \frac{7.5 \text{ hp} \times 746 \text{ watts/hp}}{1000 \text{ watts/kW}} \left[\frac{1}{0.843} - \frac{1}{0.902}\right] \times (4000 \text{ hr}) (\$.06/\text{kWh})$$

$$\text{Savings} = .434 \text{ kw } (4000 \text{ hr}) (\$.06/\text{kWh}) = \$104.16$$

There are also the accrued energy savings based on reduced branch feeder and distribution panelboard feeder line losses in the form of additional I^2R heat losses. The above line losses can be similarly analyzed as was shown under the power factor correcting section, keeping in mind it is the difference in the square of the current for the two different motors times the number of conductors times the length of run.

Table 1-1. Efficiency of General Purpose Motors.

MOTORS (HP)	EFFICIENCY RANGE (%)	AVERAGE	AVAILABLE IMPROVED EFFICIENCY MOTORS
1	68-78	73	85
5	78-81.5	80	89
10	81-88	85	90
15	84-89	87	92
25	87-90	89	92
50	88-92	90	92
75	90-92	91	93
125	91-92	92	94

Table 1-2. Cost and Efficiency of High-Efficiency Motors (Energy Efficient)—Typical Efficiency; Price Ranges.

(HP)	EFFICIENCY (%)	COST ($)
1	75-84	140-210
5	85-88.5	215-333
10	88-90	300-522
15	88-92	500-640
25	90-92	750-1,000

Thus, 3 #10 AWG conductors used to feed a 7.5 hp motor 100 feet away from its source would save the following energy and cost:

$$\left[\left(\frac{7.5 \text{ hp} \times 746 \text{ watts/hp}}{0.843 \text{ EFF.}} \div 360 \text{ volts}\right)^2 - \left(\frac{7.5 \text{ hp} \times 746 \text{ watts/hp}}{0.902 \text{ EFF.}} \div 360 \text{ volts}\right)^2\right]$$

$$\times \text{ (3 conductors) (100 ft) (0.1018 ohm/100 ft)}$$
$$= [(18.436 \text{ amps})^2 - (17.230 \text{ amps})^2] \text{ (3 conductors) (0.1018 ohm)}$$
$$= (339.89129 \text{ amps}^2 - 296.8808 \text{ amps}^2) \text{ (0.3048 ohm)}$$
$$= (43.01 \text{ amps}^2) \text{ (0.3048 ohm)} = 13.1094 \text{ watts of energy saved.}$$

Table 1-3. General-Purpose and Energy Efficient Comparison of a 10 HP motor.

ITEM DESCRIPTION	CONVENTIONAL STANDARD MOTOR	HIGH-EFFICIENCY MOTORS		
		HIGH	HIGHER	HIGHEST
FIRST COST	$180.00	$224.00	$252.00	$279.00
EFFICIENCY (PERCENT)	81.8	85.2	86.8	87.4
ELECTRIC ENERGY REQUIRED (KW)	9.12	8.76	8.59	8.54
HOURS OF USE PER YEAR	4,000	4,000	4,000	4,000
YEARLY POWER CONSUMED (KWH)	36,479	35,023	34,378	34,142
COST OF POWER PER KWH	$0.06	$0.06	$0.06	$0.06
ANNUAL ELECTRIC COST	$2188.75	$2101.41	$2062.67	$2048.51
ANNUAL COST OF OWNERSHIP BASED ON AN 8 YEAR LIFE	$22.50	$28.00	$31.50	$34.88
TOTAL ANNUAL COST OF OWNING AND OPERATIING	$2,211.25	$2,129.41	$2,094.17	$2,083.39
DIFFERENCE IN ELECTRIC COSTS	$0.00	$87.34	$126.08	$140.24
DIFFERENCE IN FIRST COST OF MOTOR	$0.00	$44.00	$72.00	$99.00
PAY BACK PERIOD IN YEARS	NONE	.50	.57	.71

For the same 4000 hours that the motor is running it will produce the following line-loss cost savings:

$$\text{Annual branch feeder savings} = \frac{\text{watts saved}}{1000 \text{ watts/kW}} (\text{Hrs. of use})$$

$$(\text{cost/kWh}) = \left(\frac{13.1094 \text{ watts}}{1000 \text{ watts/kWh}}\right) (4000 \text{ hrs}) (\$.06/\text{kWh})$$

Annual branch feeder savings = \$3.15 as an additional spin-off saving along with another \$1.68 in attendant A.C. energy savings.
(Using a higher efficiency motor)

Since these higher efficiency motors are cost-effective with quick payback and large accrued annual energy and cost savings it allows engineers and owners to be able to afford a *design to energy efficiency* ethic.

This is especially true when you consider that 60% of all domestically generated electric energy is used to power electric motors. About 25% of all domestic usage is by the more predominantly used 1 through 125 hp AC polyphase motors used in industry. Although they comprise only about 20% of all motors in use, their energy usage accounts for a staggering consumption 16 times greater than the remaining 80% of the motors mainly in the fractional hp range.

CONCLUSION

The incorporation of these techniques into a carefully engineered overall design will result in efficient, energy conservative buildings and distribution systems. The quantities of finished electrical hardware used in a given project will be reduced (hence, also the total energy expended to provide the building's electrical components) without any sacrifice in performance and reliability.

During the life-cycle use of a building, less energy will be used since less material will have to be replaced and maintained. Most importantly, less energy will be required in the support of the building systems since they will be less energy wasteful and more energy efficient. Each subsystem will be more energy efficient and will impart energy efficiency to other disciplines by reducing impact loads on these other systems, such as air conditioning.

2
APPLYING INTERDISCIPLINARY ENGINEERING AND ARCHITECTURAL VALUE ENGINEERING ANALYSIS FOR MAXIMIZING ENERGY CONSERVATION

EVALUATION OF INTERDISCIPLINARY INTERACTION

To reduce the first cost of any construction project, items which are the most costly must be identified, given priority attention, and costed out. Each discipline must evaluate its own design characteristics and constraints to ascertain the cost/worth ratio of all its systems and subsystems. These design decisions must be weighed in terms of how they in turn impact other disciplines. For example, how does the design decision to save 10% of a specific element affect other systems? Will the decision to change that element incur a greater increase in cost to one or more of the other trades directly affected by it?

RESIDENTIAL FURNACE ENERGY-CONSERVATION TECHNIQUES

For residential buildings, the conventional design approach is to use a centrally located fossil fuel (gas or oil) furnace which usually totally negates the energy savings, cost/worth ratio, net gain, or accrued value of installing double-glazed windows or adding additional thermal insulation.

The furnace, while heating the residence by forced air induction, heats the surrounding space by convection and radiation. Air is drawn into the furnace by an induced draft which uses the air for combustion and also sweeps it, along with additional heated air, up the flue. Cold outside air must seep in through cracks or leaks in the structure to replace the air for combustion. That air is also swept up and lost through the flue. The loss is thus compounded in that surrounding air is heated and then lost to combustion or through the flue (heated air is less dense and less efficient for the support of combustion), and cold outside air which must also be heated is drawn in to replace it. Caulking double- or triple-glazing (storm windows as well) will reduce outside air infiltration and reduce the heating burden, but if air is being drawn out of the house through the furnace, then additional air will need to be drawn in to replace it. Thus, caulking and glazing would be counter-productive in this instance and should not be used.

A centrally located and installed furnace is considered lower in first cost (for its material and installation) but costs the owner more in fuel bills. If, however, colder, denser outside air is ducted into the furnace for more efficient combustion, then the already heated household air is not lost and additional heating energy is not expended. In this instance, caulking and glazing will be beneficial and will not only save the future owner money in lower fuel bills by lowering energy waste but will reduce first costs by allowing a smaller furnace since the heating requirements will be lower by virtue of lower heat losses. Thus, an existing installation can be converted to a more energy (fuel) efficient house by the addition of an outside-air, combustion-air duct. As a result, whether you are contemplating or have specified additional caulking, storm windows, etc., or you already have designed or installed these energy-saving features, they will become even more effective and energy efficient.

If you are presently considering replacing an existing furnace with a more fuel-efficient one, you can also take advantage of using a still smaller and even more energy-efficient unit. Energy-efficient furnaces utilize automatic control dampers in the flue and a solid-state electronic pilot system. (See Figure 2–1.) The damper control prevents heated air from being drawn up the flue when the heating plant is shut down, which is ideal for a non-ducted outside air system but is still not as efficient as the ducted system. The electronic pilot system (in lieu of a gas pilot light burning all year long as well as in between furnace operations) may save a household owner up to 9 dollars a month. The above energy savings analysis also applies equally well to fossil-fuel (gas or oil), domestic, hot-water heaters. The same approaches and techniques can be applied to them as well, but since they consume less energy in the home than the central furnace, the energy saved will also be less.

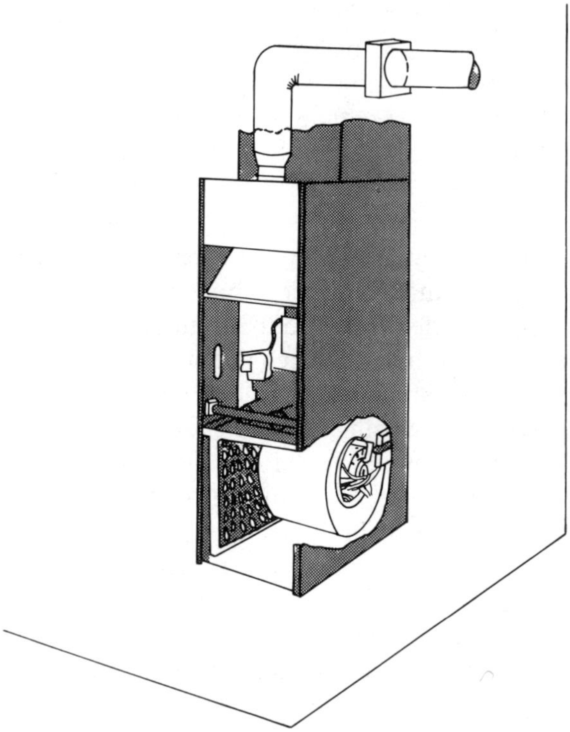

Figure 2–1. Centrally ducted furnace.

REDUCED FLOOR-TO-FLOOR-HEIGHT COST SAVINGS

One very cost effective technique for reducing first costs is the reduction of the floor-to-floor height. This can be accomplished only if the necessary access space for utilities is not constricted to the point that the cost of the other disciplines is increased disproportionately due to severe space restrictions and coordination difficulties. By reducing the floor-to-floor height, the cubage associated with the overall floor area is reduced. Two advantages result and both ultimately lead to first-cost savings: 1) less cubic feet per unit area, and 2) a reduction in perimeter curtain wall or masonry wall per unit area.

In Washington, DC there is a building height restriction to preclude obscuring (by virtue of sheer size and height) various federal structures and monuments, notably the Capitol and the Washington Monument, to name but two. The total height restriction for district C-5 (the Pennsylvania Avenue Development) is 130 feet. This will allow a 10-floor structure having a 13-foot floor-to-floor (slab-to-slab) height, or 12 floors at 10′10″ floor-to-floor. In the second instance there is more leasable space for nearly the same or slightly more money, giving you more income for your bucks. The cost/worth is such that there is far more worth of goods and service to be accrued. than the cost incurred, as well as little additional energy expended in constructing and operating the larger floor-space facility.

Another advantage of lower floor-to-floor heights is that ceilings in general are lower. With a 10′10″ slab-to-slab height, the ceiling height will probably be 8′0″ above the finished floor (A.F.F.) to no more than 8′6″ A.F.F., although there will be a decreased ceiling cavity space for heating, ventilating and air conditioning (HVAC) ducts and wiring. A reasonably lower ceiling height of around 8′0″ to 8′6″ has a twofold advantage. With a lower ceiling height the first advantage is that a greater dropped ceiling cavity space is available to run utilities unobstructedly, thus facilitating their ease of design and installation. This, in turn, reduces the cost of both designing the facility and building it, as well as reduces the energy expended in its construction. The second advantage of a lower ceiling is that less energy is needed for lighting to a specific illumination level (for the same light source and luminaire). As the ceiling height is increased, the energy requirements will increase to sustain the same lighting level. According to the inverse square law for a point source, when the distance is doubled the lumens have to be quadrupled to obtain the same number of footcandles on a given surface. Thus the energy consumption is increased fourfold.

Since fluorescent sources and their attendant fixtures are normally used, and they are not point sources but linear and area light emitters, a doubling of the ceiling height (and hence the height of the lighting fixtures above the work plane which is more than doubled) does not require quite as much as a fourfold increase in energy. However, incandescent and HID-source downlight luminaires are essentially point sources. They would indeed increase their energy needs a little more than four times over to obtain the same lighting level for a doubling of the ceiling height since the mounting height above the same work plane will be slightly greater than doubled. (See Figure 2–2.)

The value engineer and the design professional must be wary of not boxing themselves into a corner. Obviously a reduced roof height or a reduced floor-to-floor height in conjunction with a constant ceiling height will produce a reduced ceiling cavity height (space) in direct proportion. This in turn requires the solution of additional problems inherent in the coordination of structural, HVAC, and lighting (electrical) disciplines and trades, in an increasingly tighter ceiling cavity space. With these constraints the architect, working in conjunction with the electrical, mechanical and structural engineer, may have to direct the structural engineer to design (or use) a structure of the lowest floor support depth possible.

Costs of structural construction have to be evaluated and compared to other building costs. Will a structural discipline's higher first cost in construction of a flat plate (shallow depth) design versus a lower first-cost structural joist supporting concrete on a steel deck (with its inherent greater depth and space restriction) reduce the overall cost of all disciplines involved? These coordination problems can sometimes be increased for an integrated ceiling system. If properly designed and coordinated, an integrated ceiling system can however reduce field installation problems.

PRECAST, PREWIRED JUNCTION BOXES

There can be little or no economic advantage or justification in precasting conduits and junction boxes (JBs) in the ceiling slab and prewiring them for future tenant lighting and receptacle circuitry

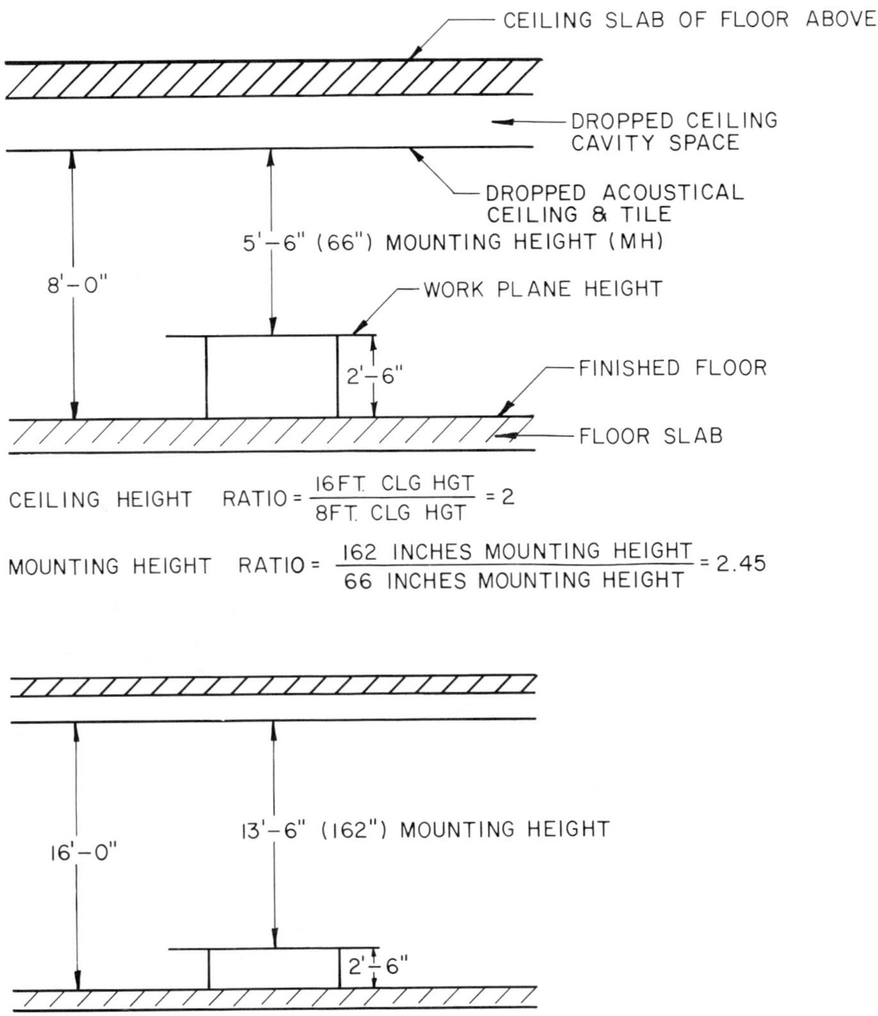

Figure 2–2. Ceiling height versus work-plane height relationship.

work as is being done in general practice for speculative office designs. If the provisions are inadequate for the project design, then additional branch circuitry will have to be brought in at additional expense. If too much spare capacity is there, then that additional money is not utilized and is wasted. It is much better to tailor the design to the needs of the project at the time of the actual tenant design layout or installation.

Armor interlocked cable has a limitation in that it can only be installed in non-environmental air cavity spaces or be used in dropped ceiling cavity spaces used for environmental air if in lengths of four feet or less. However, flexible tubing (greenfield) in lengths of 6 feet or less or liquid-tight, flexible, metallic conduit in single lengths not exceeding 6 feet and electrical metallic tubing (EMT) with conductors can be run in the environmental-air, dropped-ceiling cavity space. When supply and return air is ducted, then prewired metallic sheathed cable, such as BX, can be run in the ceiling cavity. BX can be run in all tenant partition walls for receptacle circuitry. BX, greenfield or EMT can be drawn into the ceiling space later, as required, and for less money than a concrete precast, prewired JB system.

An added advantage is that it negates inordinate oversizing of branch circuits (both in numbers of wires and conduit size) in attempting to provide adequate spare future capacity. Thus underloading of circuits is eliminated and, in addition, the need for additional branch circuit conductors, circuit breakers and light switches for local control is decreased. However, more light switches may be required since at least one switch is required for each phase wire or branch circuit. By loading the branch circuitry to full capacity, perhaps 50% less branch circuits would be required, thus cutting in half the number of switches and switching circuitry.

The precast, prewired junction box system then is the classic example of the engineer who allows a 50% capacity for future growth, which thus requires 200% more branch circuits, switches and circuit conductors.* This approach increases the energy expenditure for mining, refining and installation of more finished goods, as well as increases future energy use and maintenance costs attributed to it.

PREWIRED MOTOR CONTROL CENTERS

Motor control centers should be specified as Type II, wiring class C (Type C) construction in lieu of Class 1, NEMA Type A or B. This prewired configuration reduces first cost by minimizing installation

---

* (100% capacity/50% capacity) × 100 = 200% increase.

labor. It will also minimize life-cycle cost because of maintenance due to rehabilitations, motor changeovers or motor control circuitry changes brought about by new duty cycles. The prewired harness allows a rapid, system-wide control changeover for building automation systems (BAS) at very low connection costs. It is also an energy conservative approach in that all system components are reuseable rather than disposable due to built-in obsolescence. With BAS interconnect capability, additional operational building energy savings are also possible.

NEW SERVICE UNDER EXISTING PAVING

To minimize the disruption to existing paving, excavation and backfilling, facilitate the installation of new gas, water, telephone and electric services, and reduce costs, the use of a hydraulic ram rather than using cut-and-cover techniques is far more economical. A hydraulic ram is used to bore through the earth without disturbing the area around the hydraulic ram. It eliminates the disruption and destruction of surface features such as sidewalks, streets or parking lot surfaces. A hole or excavation pit is dug at the starting and ending points and must be of sufficient size and shape to accommodate the ram. The first pit is for positioning the hydraulic ram. The oblong excavations are aligned so that upon completion the hydraulic ram can be retrieved after it emerges from the terminal excavation point. In many instances one excavation pit is all that is required since the hydraulic ram can be backed up to its original starting point. The hydraulic ramming device is known as a *Hole Hog*. (See Figure 2–3.) Care must be taken in lining up the ram's path to coincide with the raceway trench excavated on the opposite side of the paved area. After the contractor has tunneled underneath the existing paving, the various building facilities and raceways can be inserted through this opening. These pipes and raceways should be held firmly in place with plastic spacers after which insertion and entrainment of a concrete grout can take place with a slick pipe. The grout should consist of small pebbles for the aggregate which provides a thick viscous cement slurry.

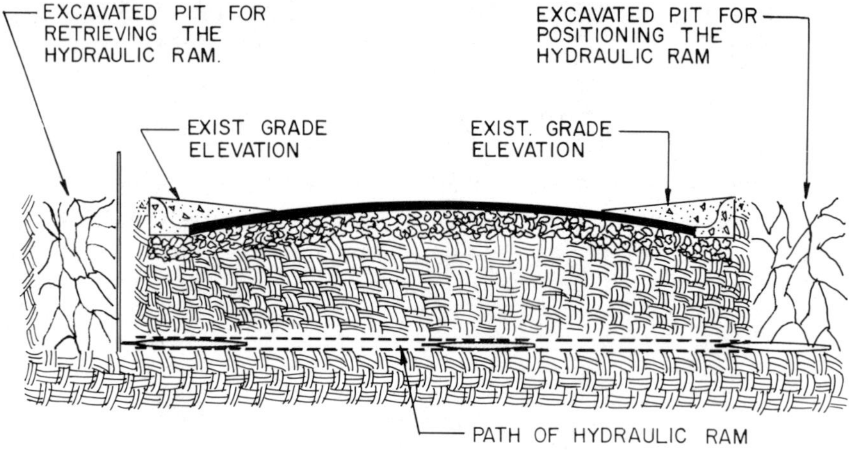

NOTE: THE CONTRACTOR TUNNELS UNDERNEATH THE EXISTING PAVING WITH THE HYDRAULIC RAM. THE ELECTRICAL RACEWAYS ARE INSERTED THRU THIS OPENING AND ARE HELD FIRMLY IN PLACE WITH PLASTIC SPACERS AND ENTRAINMENT OF A CONCRETE GROUT USING PEBBLES AS AN AGGREGATE (CEMENT SLURRY).

Figure 2–3. Hole hog detail.

EMERGENCY POWER

An emergency generator standby power system is designed expressly to provide standby electric service in the event of an electric utility outage. It is used to serve life and fire safety loads comprised of emergency white and exit lights, fire alarm and fire detection systems, possibly a public address system, telephone system, master antenna television system (if installed), and elevators. On the basis of both first and life-cycle cost savings, as well as energy savings, we should reduce the emergency generator's size by designing a system that allows elevators to be lowered to the first floor level one at a time, after which only one elevator will be in service. Standby power should be fed to the load through a transfer switch which will automatically revert to the normal utility source when its voltage and frequency return to normal.

When the utility power falters or fails, the transfer switch signals the generator to start up. When the generator reaches operational voltage and frequency, the transfer switch transfers the load to the emergency source. When normal utility service has been restored for a given select time, the transfer switch then selects that source. The switch should allow the emergency generator to run unloaded for 4 to 5 minutes before shutting down so as to reduce sludge buildup and to maintain the set in adequate serviceable condition, which will also reduce life-cycle costs due to fuel bill energy savings. The transfer switch should have an exerciser control to test the generator set once a week to keep it in operational condition. With an exerciser control, the set has a greater chance of coming on line and assuming the load in the proper time span for life safety and thus comply with the NEC. This design technique allows us to further reduce the size of the set and benefit from additional accrued energy savings. It will be in better running condition, which will further reduce life-cycle costs due to fuel bill energy savings.

To insure rapid starting, the generator must have an in-block heater to maintain coolant water within the normal operating temperature range of about 160°F to 180°F. It should not be used solely to warm the engine block during cold snaps. Utilization of municipal cold water for engine cooling should not be considered since the water at 60°F to 80°F may not come up to the full normal operating temperature of 160°F to 180°F. In addition, the added cost of the water service (and its companion sewage charge as a result of this) represents a waste of resources and energy as well as higher life-cycle costs.

A diesel engine standing at ambient room temperature without an in-block heater takes longer to start and accelerate to operating speed. This lag in starting can cause a longer delay in transfering to emergency generator power after a utility power failure. For first cost and energy saving considerations the set should not be larger than the load requires. Upon start-up it must be able to carry the full load impressed upon it, which it cannot do if it is running at below normal operating temperature. A cold start under full load can cause the generator to either drop off the line (loss of all emergency standby power) or to suffer a reduction in voltage or frequency output. Motors fed by the generator will thus experience a reduction in rotational speed during a frequency reduction. By preventing voltage reduction, the possibility of reduced lamp light output in emergency lights can be eliminated (as well as the possibility of extinguishing HID lamps), and electric motors will be maintained at full power and torque and will continue to operate within normal temperature ranges. A voltage reduction elevates the electric motor's operating temperature due to an increased current draw and additional I^2R losses in the form of heat.

The omission of an in-block heater and exerciser switch may require oversizing the generator set to assume the load initially and carry it. This causes still greater additional first cost, as well as increased life-cycle costs and increased energy consumption. Running below design temperature also contributes to acid and sludge build-up within the diesel engine crankcase.

Rather than exhaust the emergency generator at roof level or in any exposed way, the generator can be exhausted through a transite pipe into a rock-filled pit at ground level. This will eliminate expensive sleeves, chases and long vertical exhaust-pipe runs thereby reducing the first cost and energy expended to erect the facility. This system will effectively muffle the exhaust while preserving the scenic character of any landscaped grounds. The rockfill also acts as a catalyst and oxidizes carbon monoxide, converting it into harmless carbon dioxide. (See Figures 2–4 and 2–5.) A weep hole must be provided in the

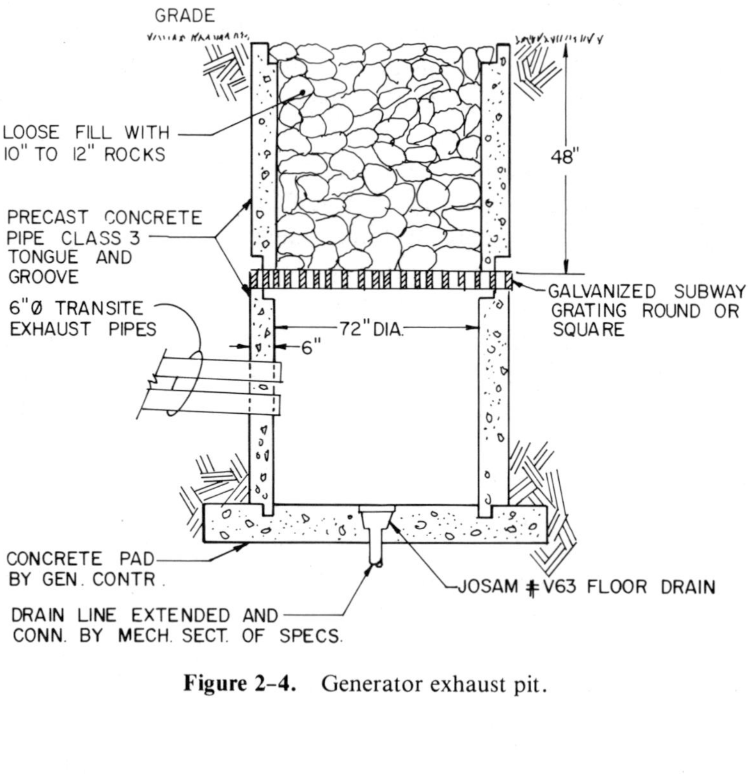

Figure 2–4. Generator exhaust pit.

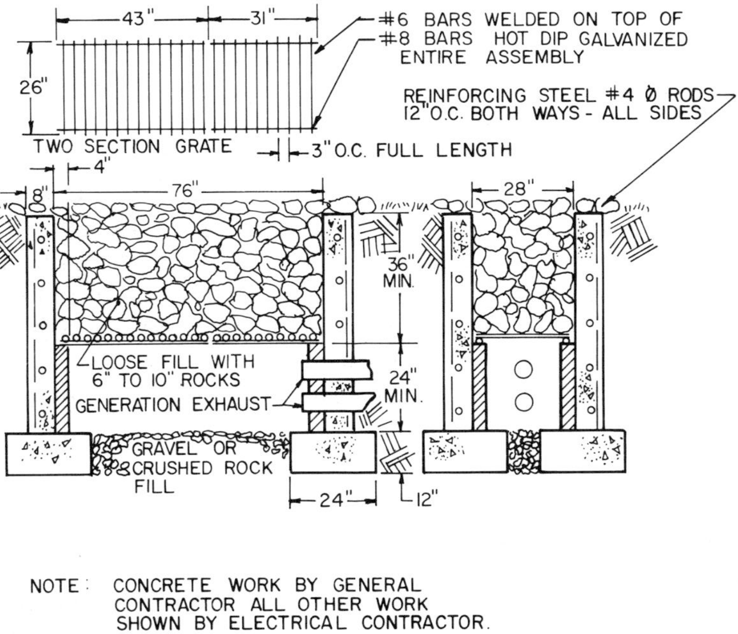

Figure 2–5. Generator exhaust pit.

exhaust pipe, close to the exhaust manifold, to drain water vapor condensation that could otherwise prematurely knock out the muffler due to corrosion. The deletion of this weep hole may cause a piston water lock that can bend the connecting rods.

Elevator motors should be supplied with reduced voltage starters (part winding) which will reduce their inrush current by about 50%; the normal inrush current is between five to six times full load amps. Normally a 25 hp motor with reduced voltage starting will require a 75 kVA generator just for start-up and support. This generator would have to be provided with resistive current suppressors to eliminate dieseling of the emergency generator during elevator descent. During the elevator descent the motor of the elevator motor-generator set acts as a generator and backfeeds the generator of the engine-generator set, which then acts as a motor and may overdrive the diesel engine.

However, by matching the diesel emergency set's output to the load and to the input curve of the elevator motor, it is possible to use a smaller generator to support life safety loads (exit and emergency lights, etc.), sequentially lowering the elevators one at a time and then maintaining one elevator in service thereafter. By thus comparing and matching of the inputs and outputs of the system components a design can be realized which permits very little voltage *droop* (a depression of voltage output, i.e., 250 volts output versus a normal 277 volts), good voltage stability (no excursion of voltage up and down, i.e., from 260 to 300 volts versus 277 volts output) and tighter frequency regulation (60 Hz) at a reduced life-cycle cost. All of the above techniques, while reducing the size and cost of the emergency generator, also reduce the amount of tool and equipment energy needed to manufacture and construct the initial installation. This in turn reduces life-cycle energy use due to the reduction of materials and equipment used, as well as reduces energy use by the engine-generator set during its useful life and use.

ALTERNATE STANDBY POWER SOURCES

When the emergency panel is ahead of the main overcurrent protective device, it becomes merely a normal power source panel, so that the loss of utility power will cause the loss of power to the emergency lighting and fire alarm systems as well.

The most economical and energy-conservative approach is to provide standby power to a facility (especially one only partially rehabilitated in certain select areas) through the use of individual battery back-up systems for the fire alarm system. The emergency white fluorescent fixtures should be equipped with integral inverter-rectifier emergency power packs. This approach will be considerably less costly than providing a new emergency generator for a building if an emergency power source is not required for the elevator. The emergency generator set and additional peripheral equipment requirements such as an attendant fuel tank, exhaust stack and intake air system can be eliminated, thus saving both material and money, as well as life-cycle savings in energy.

Generally, emergency lighting should be provided by fluorescent fixtures equipped with original equipment manufacturer's (OEM) integrally installed standby power packages within their ballast channels. Emergency white lighting fixtures would thus be maintained throughout an emergency by the standby power packages. These units combine a rectifier, floating battery source and inverter, and provide the code-specified 90 minutes of emergency illumination in the event of a power outage. Emergency lighting fixtures should be placed so that they will illuminate exit signs externally. This recommendation conforms with current codes of all local jurisdictions around the greater Washington metropolitan area, with the notable exception of Montgomery County and Washington, DC, which in the past two-and-a-half years have changed some of their code requirements. For all but Montgomery County and Washington, DC, this is an acceptable alternative to providing emergency-powered exit signs (which would be internally illuminated during an outage). This eliminates unnecessary and expensive, redundant backup equipment, thereby reducing life-cycle costs.

Fixture-installed, inverter-rectifier system, emergency power packs (available from a number of manufacturers) are alike in operating only one of the two lamps per lighting-fixture ballast and inverter/rectifier. They differ widely in the percentage of lumen output (light) at which they operate the single lamp. A comparison of comparably priced units of seven manufacturers illustrates these differences.

MANUFACTURER	PERCENT LUMEN OUTPUT
Bodine	20%
Chloride	30% to 35%
Dual-lite	20%
Exide	17%
General Electric	80%
Lightalarms	100%
Siltron	78% to 80%

If it is decided to make extensive use of high intensity discharge (HID) lighting, then HID emergency white fixtures should be powered by a centrally located inverter/rectifier.

The emergency source has to be a unit similar to an uninterruptible power source (UPS) with little power discontinuity and a little fast transfer time. UPS system are usually completely solid-state, consisting of AC to DC rectifiers, batteries floating (charging) on the line, inverters to reconstruct a series of square wave outputs into a staircase approximation of a sine wave and Pi or T filtering to smooth out the output curve.

An inverter/rectifier system for a project using extensive HID lighting has to have a logic circuit that slaves the emergency power output into the wave pattern of the normal utility source so that when the utility power falters or drops out, the system's output is in nearly exact phase with the normal utility power. This eliminates any apparent power disruption or discontinuity. The NEC requires a battery system for the inverter/rectifier system to have a minimum 90-minute standby capacity.

Unlike a UPS, an inverter/rectifier system has a temporary transient power loss either due to shifting from direct utility-power output to internal battery-power inverter output (normal design) or caused by the external (utility) service feeding the inverter DC power after passing through the rectifier, then shifting back to internal battery power to drive the inverter, which also causes a temporary power loss. For these systems to function properly for HID use, the system must transfer its power through a solid-state transfer switch in less than 4 milliseconds in order to not extinquish the arc plasma of the mercury vapor or metal halide arc tube. The cost of UPS sets is about \$1.50/W; ultrasophisticated inverter/rectifier systems cost about \$1.34/W. Using either centrally located ultrafast transfer inverter/rectifiers or using OEM installed inverter/rectifiers in fluorescent ballast channels is less costly and more energy efficient. Less materials are used and thus less tool and equipment energy is needed—an example of good source energy conservation versus boundary energy conservation. Less energy is used to maintain a charged condition; so it costs the owner less, along with reducing attendant air conditioning loads. Less energy loss also means less heat buildup in the electrical system as a whole. This decrease in I^2R equipment heat loss marginally increases the life of the electrical equipment while decreasing the life-cycle operating costs.

Central inverter/rectifiers come in square wave and sine wave outputs. Nearly all manufacturers produce a square wave output unit because of manufacturing problems with the sine wave products. The ballast of a fluorescent or HID unit neither perceives any difference, nor functions any differently with either wave form. Symbolically and perceptually we should not envision the ballast riding up and down an undulating smooth sine wave versus a bumpy, choppy and hazardous journey up and down over a series of square blocks.

One operational difference observed in a ballast served by a square wave is that it generates additional third harmonic energy losses in the form of heat that decreases its life. If inverter/rectifiers are used for standby power to emergency white lighting fixtures, they should have the utility power feed the ballast under normal conditions. Then only under utility loss will it be subjected to additional heat and accelerated aging.

An emergency generator set as a source of standby power for emergency white HID lighting fixtures can meet code requirements but still seriously endanger the occupants of the building. The arc plasmas will be extinguished in a MV, MH or HPS lamp while waiting 5 to 10 seconds for the emergency generator set to come on line. A 10- to 15-minute wait for MV or MH is then required for the HID lamps to cool down before restriking. An additional 8 to 10 minutes are required for the lamps to come to full intensity. HPS and high color rendering index (CRI) HPS restrike in 1 minute after extinguishment and produce approximately 30% intensity immediately. They restrike immediately after extinguishment if dimmed down by special dimming equipment to 48% or less of energy use and light output. They then reach 80% of full intensity within 5 to 8 minutes.

SOLID-STATE, VARIABLE VOLTAGE, VARIABLE FREQUENCY MOTOR STARTERS

An analysis of the reliability of most electric utility companies indicates that they have inherent power utility abnormalities of the following types:

Impulse—any high frequency transient increase or decrease in line voltage lasting for less than one-half a cycle; *sag*—a decrease in line voltage lasting one-half a cycle or more; *surge*—an increase in line voltage lasting one-half a cycle or longer; *brownout*—a programmed reduction in the voltage level by the utility company lasting for hours or days; and *blackout*—any complete loss of power.

Computers and other sensitive, highly sophisticated electrical and electronic equipment are susceptible to these abnormalities. Because of the frequency of these abnormalities, often referred to as *glitches* or *high-speed spikes* (as well as the possibility of brownouts and blackouts), it is usually advisable to provide some form of standby power for emergency HID lighting fixtures and sensitive computer equipment that is not equipped with a hard core memory. However, rarely do fan and pump motors experience electrical or mechanical difficulties due to utility abnormalities.

When an electric utility has a transient loss of power, an air conditioning compressor's motor, fan motor or pump motor may well act as a generator. The back electromotive force (emf) that is generated when out of phase with the utility power coming back on line can cause a braking effect or transient torque (reverse torque) problem. A reverse torque in a large pump or chiller compressor motor may be enough to snap the interconnecting shaft or shear the coupling flange bolts. 50 milliseconds or $\frac{1}{20}$ of a second represents only a 3-cycle duration of a 60 cycle per second or 60 Hertz power source. 100 milliseconds being twice as long represents $\frac{1}{10}$ of a second or a 6-cycle duration. Most air conditioning chiller compressor motors are 2-pole devices having a nearly 3600 rpm rotational speed. Air handler fans and pump motors quite often come supplied with 1725 rpm motors.

A 4-pole motor with two pairs of poles produces 1800 rpm (60 CPS times 60 s/min divided by 2 pairs of poles equals 3600 divided by 2 equals 1800 rpm). Allowing for slippage, in order to induce a current in the rotor by cutting the magnet flux, an actual speed of about 1725 rpm is attained in ordinary practice under full-load conditions. When only a 50 ms duration outage is incurred ($\frac{1}{20}$ s), the motor has only rotated 1.4 revolutions (1725 rpm $\times$ $\frac{1}{20}$ S) and has not yet had sufficient time to build up a substantial back emf capable of bucking the utility voltage when it comes back on the line.

The only immediate transient impulse or *voltage spike* occurs at the instant of voltage decay (open circuited condition) to the primary windings of a utility step-down transformer or to the motor windings, which produces a high potential voltage but relatively little current. Real power equals voltage times current in phase with voltage: $P = VI\cos\theta$. Since the current is nearly zero, the real power to create a sufficient transient torque problem is not available.

The effect of voltage decay on a motor is similar to conditions in the ignition system of an automobile. When the ignition points close, current rushes into the ignition coil and encounters a large resistive force (both resistive but primarily inductive reactance). The coil acts as a choke producing a back emf which resists the voltage impressed upon the coil. As the current increases, the magnetic flux or *gauss* also increases. As the flux increases and spreads out like a shock wave it crosses the copper wound coils of the secondary winding of the ignition coil producing a high voltage (about 15,000 volts). (See Figure 2–6.) The electric spark cannot jump the spark plug gap due to the

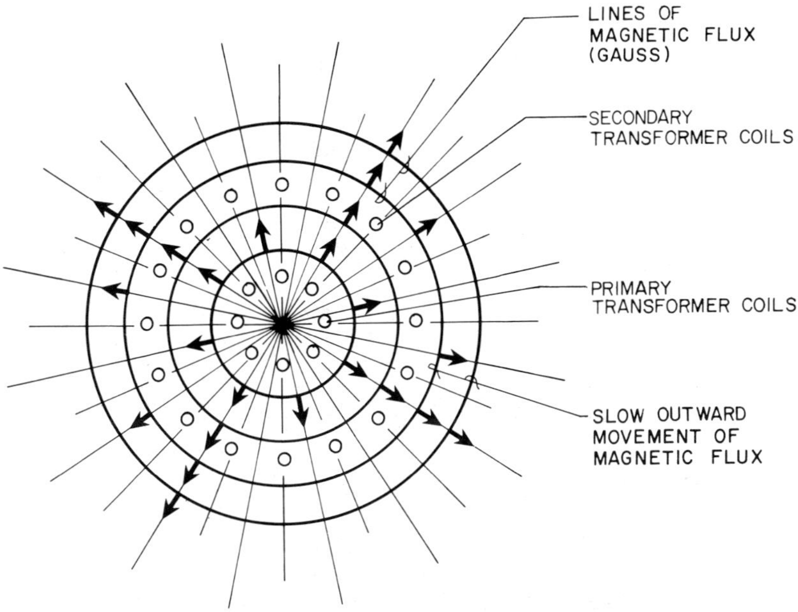

Figure 2-6. Magnetic induced voltage. Primary stage induction of voltage in the secondary coils.

high resistance of the air. As the coil's primary current reaches a maximum, the secondary coils reach saturation (maximum number of magnetic lines cutting the secondary coils). At this point, no outward lines of flux are produced and the secondary coil's voltage drops to zero. In most instances, the ignition points would have already opened, causing the loss of all voltage impressed on the primary coil's terminals. This results in an immediate reduction of the current to zero, producing an immediate collapse of the magnetic lines of flux. This inward collapse of flux (similar to a pebble being tossed into the center of a pan and then having its ripples bounce off the rim of the basin and back to the center), causes a much higher voltage to be induced for spark plug ignition. (See Figure 2-7.) This, in effect, is the immediate type of transient voltage (capable of little or no work function) that is produced in the motor.

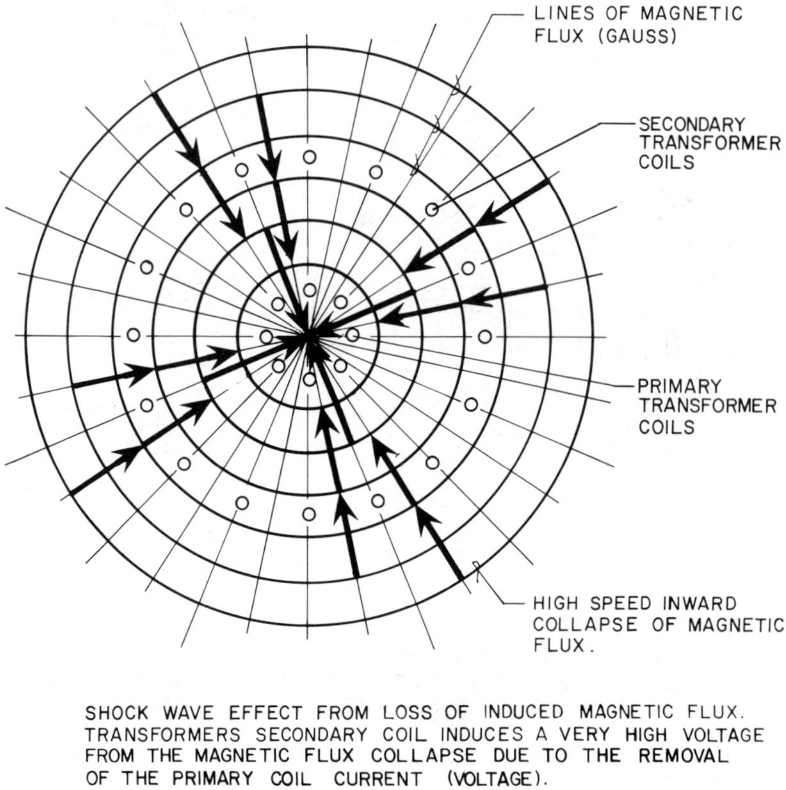

Figure 2–7. Magnetic induced voltage. Secondary stage induction of voltage in the secondary coils.

On the other hand, a longer duration power outage of over 50 ms should be sufficient time for the control circuit coil's paralleled contactor and the motor contactor operator coil to drop off the line, thereby precluding the motor's back emf and the utility's emf, producing a reverse torque condition. (See Figure 2–8.) It is conceivable, however, that the reverse or back emf might be of sufficient magnitude (energizing capability) within a short enough time frame to cause the holding coil to hold on line. However, if this condition were to occur and a transient "reversed" torque problem due to the two sources (back emf and utility secondary service) being out of phase did exist, the motor's molded case-feeder circuit breaker should sense the short-circuit condition and thus its magnetically actuated armature

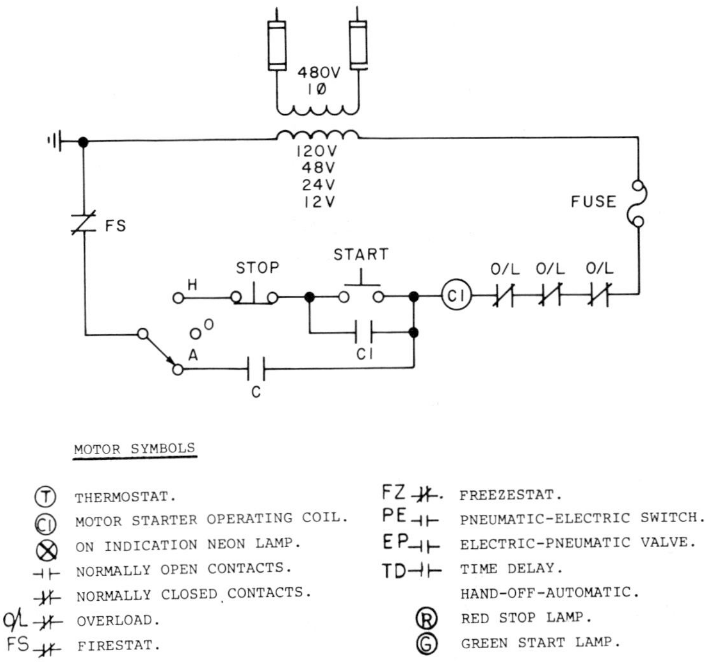

Figure 2–8. Motor control diagram.

would open-circuit it. Many intermediate polyphase (three phase) motors from 1/2 to 75 or so horsepower are hard started across-the-line. That is, the electric motor is started by applying full across-the-line, line voltage, be it 208 volts or 480 volts.

Many large chiller motors have either open or closed transition star delta starters that start the motor in an "apparent" reduced voltage mode at either 120 or 277 volts wye-connected configuration and then make a closed transient change to a higher operating voltage configuration of either 208 volts or 480 volts delta. (See Figure 2–9(A), 2–9(B) and 2–9(C).) This technique is used to reduce the inrush current. The inrush current is equal to the locked rotor current and as a rule this is usually six times the full load current. Most utility companies have clauses in their tariffs limiting the inrush kilovolt amp KVA (current) to no more than 1/6 of the total connected KVA or transformer size.

In general, any high speed spike on the primary side of the transformer would tend to be masked out on the secondary side of the coils.

A transient voltage loss of 5 to 10 ms on the primary side would be seen as a high-speed voltage spike or transient impulse on the secondary side. So, any power loss on the primary side of 5 to 10 ms should not cause any problems to a large motor such as a chiller, other than a low-grade, transient, thermal shock of no relevant importance.

Losses on the primary and secondary side of the transformer would normally last approximately $\frac{1}{2}$ second since most primary circuit breaker reclosers require about 26 cycles for operation (approximately $\frac{1}{2}$ second) to reclose.

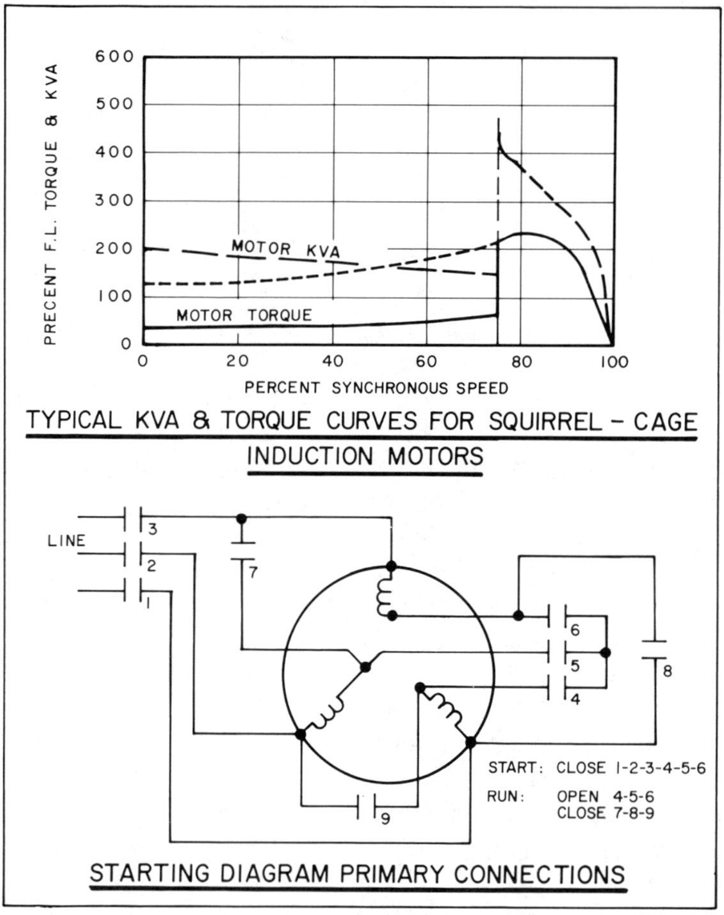

Figure 2–9(A). Open transition wye-delta (star delta) starting method.

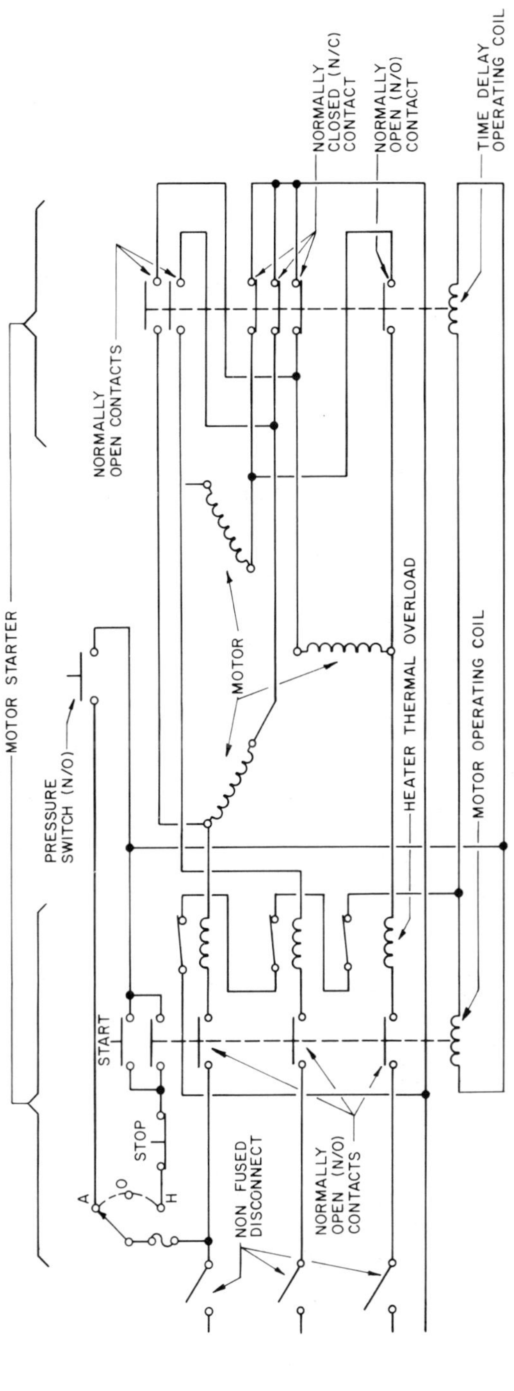

Figure 2–9(B). Reduced voltage wye-delta motor starter.

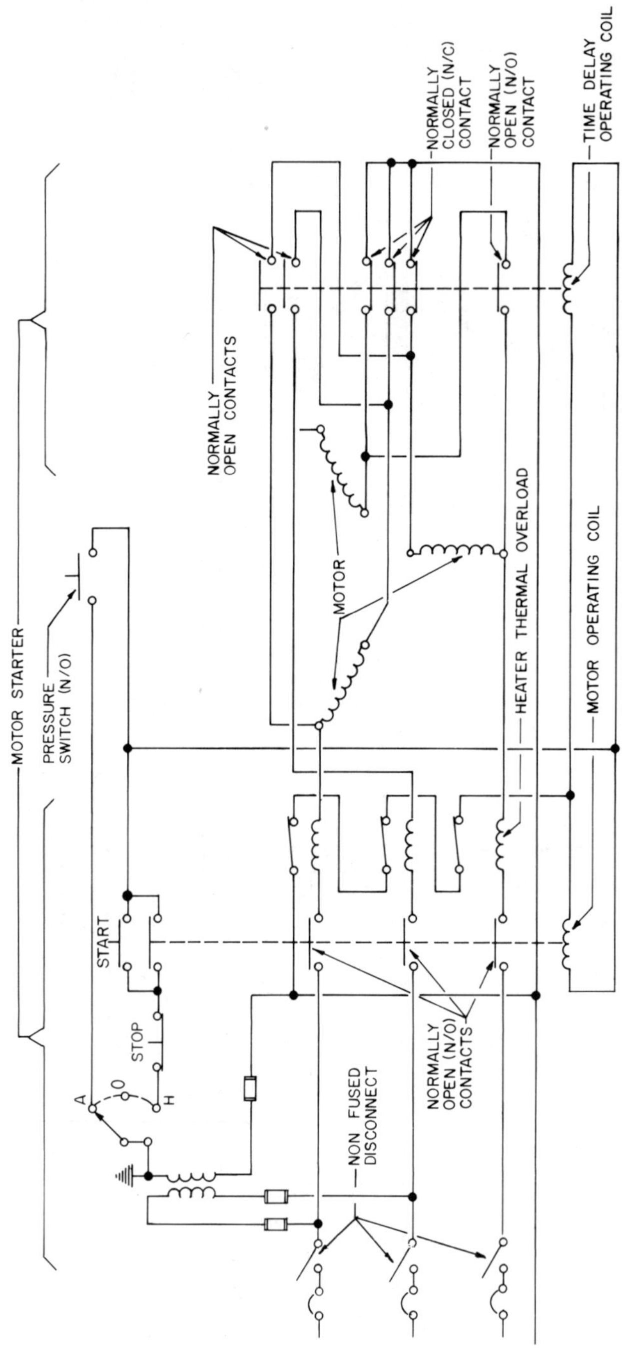

Figure 2-9(C). Reduced voltage wye-delta motor starter.

A final thought should be given to transients as they relate to open-transition, reduced-voltage, star-delta (wye-delta) starter operation. In the open transition mode, during the transition from the star to the delta connection, the contactors may create anywhere from a 50 to 100 millisecond pause or discontinuity. This raises the question that if the starter produces this type of transient outage (albeit at a much lower rpm during start-up), why should this be so much less of a problem than a transient loss of power at full rpm caused by the utility, again inducing increased energy waste? (See Figure 2–9.) Granted that the difference in rpm from start-up to full rotational speed will cause a difference in back emf, but the difference is one of magnitude. In the full rpm transient mode, the back emf is small to begin with; while during the transient mode start-up, it is even smaller.

As a result of the problems previously enumerated, it would seem good engineering judgment and practice to replace the common electromagnetic manual motor starter equipment with the newer technology solid-state starter types. These starters normally come factory supplied with 3-phase voltage phase sensors (current guard) to detect any line (phase) loss or substantial line abnormalities in order to remove the utility voltage from the motor terminals within 20 to 30 ms. Phase loss or other severe abnormalities can be detected within $\frac{1}{4}$ cycle of the disturbance, or 4.17 ms. This would normally eliminate concern about the electric utilities' abnormalities and can increase the life of the inductive motor equipment, thus increasing tool and equipment manufacturing energy savings. Line abnormalities and transients are causes of additional energy losses.

Another benefit in using solid-state starter technology in conjunction with variable frequency output and variable voltage output such as a Turbo Modulator (by York) is that triacks can be brought on from full retard to full advance gradually, thereby increasing the voltage impressed upon the motor for a soft start. Starting the machine across the line from 0 to 208 volts or from 0 to 480 volts in a gradual increase is easier on the machine than slapping the voltage across the motor terminals at either 120 or 277 volts and then hitting it again in the second and final stage at either 208 or 480 volts. This not only increases the life of the equipment and decreases energy wastage but decreases life-cycle costs.

The newer type and latest advanced technology solid-state motor starters can vary the output frequency and thus the rotational speed of the polyphase electrical squirrel cage motor.

A variable frequency output of 40 Hz will produce 1,200 rpm in a 4-pole motor and 2400 rpm in a 2-pole chiller motor. This feature, in conjunction with a chiller machine's ability to vary the pitch of its vanes for unloading the unit, allows the system to run partially loaded without paying an added energy penalty. What this means to the owner or operator of a facility having a turbo modulator installed on his chiller equipment is that it can run at half load (chiller, chilled water output) and consume no more than half the energy it uses when running at full load.

0.7 kW is required per ton of chiller capacity at full load and less than 0.35 kW is required per ton at 50% of full load. Further justification for use of this equipment is that a 30% savings per year may be expected, according to York, the manufacturers of the Turbo Modulator.

A variable frequency, variable voltage-output, turbo, solid-state motor starter system similar to York's Turbo Modulator can overcome problems of transient power losses, sags and surges and reverse torque and can also act to tailor the energy use to the controlled chilled water output, thus saving energy.

An analysis of the performance and characteristics of some of the more common types of motors will quickly show why a variable voltage and frequency, solid-state motor starter is advisable for energy conservation. A standard 2 hp drip-proof motor has the following characteristics:

- A PF of 77% and an efficiency of 80% at full capacity.
- A PF of 55% and an efficiency rate of 75% at 50% of full load capacity.

The same type of motor in a $7\frac{1}{2}$ hp size has:

- A PF of 83% and an efficiency of 82% at 100% capacity loading.
- A PF of 66% and an efficiency of 81% at 50% of full load capacity.

As can be seen, when the squirrel cage motor is unloaded, its efficiency and PF decrease. These, therefore, are the reasons that argue for the new energy-saving techniques and technologies to be applied and used.

The Turbo Modulator by York and similar manufacturers of variable frequency, variable voltage-output, solid-state starters offers additional extra cost accessories. These technological devices are designed to forewarn of preeminent failure in the chiller compressor motor or near-term potential failure in any squirrel-cage, alternating-current, electric-inductive motor. Large air handler motors, chiller motors and other large motors that must operate continuously will likely necessitate more frequent repair; thus regular maintenance is both necessary and unavoidable.

Solid-state motor starters are available with extra cost maintenance and safety features. They have bearing temperature, eccentricity and stress meter sensors which make it possible to predict failure, hours and sometimes days, in advance.

These extra cost accessories come in both local and remote alarm capabilities. Remote alarm is accomplished by transmission of a signal over a dedicated or leased telephone line to a service facility that has 24-hour-a-day, 7-day-a-week surveillance and maintenance capabilities. The potentially failing unit is thus removed from the line for servicing before a major failure or breakdown, minimizing the extent of damage and possibly minimizing the cost of service and repair. The repair mechanic will know before leaving his shop what specific item or items of equipment need replacement. What this means to the owner of such a facility utilizing this type of maintenance telemetry equipment is that he does not have to pay twice for the same repair; namely, the mechanic does not have to be deployed to the facility to check out the failed piece of equipment and find the problem, extent of damage and the replacement parts required. After such a visit he must return to his shop for parts and then return to the facility to repair the damaged motor. He may even have to completely replace it with a new motor. With telemetry monitoring capability none of the above is required. This conserves on precious mined and refined materials to produce finished goods such as the chiller compressor equipment, as well as conserves the energy required to produce these finished services and goods. This, then, provides good source energy conservation and reduces lost time on the line due to equipment failures.

A transmitted signal from a large number of subscribers to a large central monitoring and repair shop makes it economical for the shop to have in stock all the types of replacement parts required, and in quantities large enough to avoid running out of replacement parts in between inventory and restocking.

Today's computerized cash registers and accounting or invoicing machines regularly and automatically reorder and restock while billing the customer. What this means to the owner of a facility is a low cost, quick repair with minimum downtime and at the lowest energy loss. Energy loss is decreased in that there is less energy expenditure for additional finished replacement equipment since less will be needed. There will be less fuel consumption to transport goods and maintenance men back and forth, and less energy loss from continued on-the-line use of failing, inefficiently running equipment.

CONCLUSION

The incorporation of these considerations into a careful, interdisciplinary, integrated, architecturally engineered design will result in energy-efficient buildings. They will be low in first cost as well as low in life-cycle cost because of reductions in maintenance and energy use. Energy use will be reduced through the design and use of highly energy-efficient subsystems coordinated and integrated into a working entity that is even more efficient than the sum of its parts. The quantities of finished materials and hardware will be reduced, thereby reducing the total cost and energy expended. The building will provide power systems at the least cost without any sacrifice in performance and reliability. That is good value engineering and good energy management.

During the life cycle of the building, less cost will be incurred and less energy will be used since less material will have to be replaced and maintained. Most importantly, lower operation cost and lower energy use will be required to support the building's systems since they will be less maintenance prone, less wasteful and more energy efficient. Each subsystem will be less costly to maintain and will be more energy efficient and will impart cost savings and energy efficiency to the other disciplines by reducing impact factors and loads to the other systems.

3
DEMAND CONTROL OF BUILDING ELECTRICAL AND HVAC SYSTEMS

INTRODUCTION

Modern buildings have grown in complexity to the point that piecemeal, independent operation and supervision of building systems are no longer adequate. Modern building systems are characterized by their interrelationships and interdependence, and by their sheer number and sophistication.

A central monitoring and control system for electrical demand, lighting, HVAC, security, elevators, cooking, computers, life safety and other systems is fast becoming an essential capability. Its importance arises from two factors: the need for overall building energy management, and the need for dependability in the proper functioning of life safety systems during an emergency (and this, as will be explained further, is a secondary facet of energy management). A building automation system can provide this monitoring capability and control.

BUILDING ENERGY MANAGEMENT

Most buildings' energy requirements will vary seasonally, daily and even hourly because weather and occupancy influence cooling, heating and lighting needs. Energy intensive functions such as cooling, computer work, printing, conveying and heavy use of freight elevators can also have a major impact upon the building energy profile.

Although energy conservation is sound management policy in every respect, it becomes of prime importance in relation to utility demand charges. Like building electrical systems, utility systems must be sized not for the long-term average load, but for the maximum load which can occur, however brief it may be.

This maximum load may be 30% higher than the average load, resulting in the utility company having to have on the line 30% greater generator and distribution capacity than is required during most normal daily operations. For this reason, most utilities assess customers using 1,000 kW or more a demand charge over and above the normal kilowatt-hour charge when their demand rises above average levels during the on-peak hours, such as 7 A.M. to 10 P.M. In general practice, a peak demand charge by utilities is based upon the highest level of electrical energy use during a 30-minute period within that given billing period. This charge is applied to the utility plant expansion necessary to compensate for high peak demand loads of short duration.

The BAS can minimize these demand charges by regulating building energy use within preset guidelines with the assistance of a computer memory to anticipate periods of peak demand based on prior building history. A BAS can continuously monitor the electric demand meter and when necessary shed non-essential loads to reduce extra electric demand charges. Energy management and conservation is of prime importance in relation to demand charges. Some of the specific programs under the control of a building automation system can be:

- Status and alarm scanning.
- Time-based program control.
- Event-initiated program control.
- Output display commands to peripheral devices, i.e., a printer, operator's terminal and annunciators, etc.
- Record energy consumption.
- Security control.

The programs are able to measure temperature, efficiency and performance of chillers; compute weather data (degree days); and prepare energy profiles. The outdoor minimum air damper's control program can permit damper closing during unoccupied hours of the winter to retain building heat.

During the morning warm-up cycle, the dampers can and should stay closed while the building is brought up to comfort levels and only then should they be opened to bring in the required amount of fresh air just before the occupants are due to arrive.

The program can start and stop the equipment on a duty cycle (time on and time off) when the building is occupied based on temperature levels within the building (i.e., duty-cycle the equipment during early morning, lunch period and in the afternoon). The program should have the capability of allowing the space conditions to drift to a point just before the comfort levels are affected.

A computerized BAS can coordinate all automatic and operator-controlled functions, schedule functions, allocate resources, establish a building operation history, function as a memory, interface different systems and peripheral devices, and handle interruptions. The BAS can establish priorities for all functions to cut back automatically or shed functions (on a predetermined priority basis) to reduce total building energy demand before anticipated peaks are realized in practice. This is the heart of energy management. In practice, it works in many ways.

With regard to HVAC, a BAS can monitor outside air temperature and moisture content and select by automatic damper adjustment more cool, dry air for "free cooling" or only a minimum of warm, wet air to reduce chiller loads. Free cooling would be used only when all the building heating requirements obtained from the chillers double bundle heat extraction system have been satisfied and hot water temperature in the storage tanks, if used, has reached a predetermined setpoint. At that time, the chillers can be unloaded.

Temperature and humidity sensors in this type of system can continuously feed information to the computer. (See Figure 3–1.) When the outdoor air is sensed to contain the least total heat content, the computer can be used to open the outdoor air dampers, allowing chillers to be unloaded.

The BAS can also be programmed for start/stop operation of HVAC systems based on time of occupancy and change in outdoor temperature to reduce energy consumption. An operator override can also be provided. Building occupancy data such as morning start-ups, evening shutdown times, janitorial schedules and special weekends

Figure 3–1. A central building automation BAS system installed in Martin Marietta's Corporate Headquarters.

and holiday hours can be programmed into the BAS to insure that equipment is used only when and where required.

A seven-day program can also allow a different on/off schedule for each day of the week or weekend, and a holiday program with an advance preset feature can allow the establishment of on/off holiday schedules. An optimum start/stop program in conjunction with the seven-day program can save large amounts of energy each year. On average design days, which comprise the majority of days, HVAC equipment will be started later and shut down earlier to save energy; and the program can be automated. On other than the normal design days, the program can automatically delay or speed up by adjustment of the equipment start/stop time. Toward the end of the day, the program will shut down equipment based on outside and indoor conditions. Equipment will be shut down much earlier on a mild day, allowing the indoor temperatures to coast for longer periods in selected zones. Also, various zones of a building can be started or stopped separately. During the late afternoons in winter the system will keep the heat on a little longer in the cooler, eastern side of a building. It will do just the opposite with cooling in the summer.

Multiple load cycling programs can be run simultaneously in the building. This can allow different zone by zone temperature compensation limits. For example, tighter limits may be required in a data processing area and much broader ranges in general office areas. The flexibility of the program can have the ability to save energy in most zones while allowing higher loading in the other zones. As an alternative, lighting zones can be established and programmed into the system. The lighting can be controlled like HVAC, for maximum usefulness at minimum energy use. For example, with the use of automatic dimming systems, lighting levels can be automatically reduced prior to peak summer air conditioning hours, both to reduce the lighting heat introduced into the building and to offset the increase in air conditioning energy required to counteract it as well as outside heat.

EMERGENCY LOAD SHEDDING

A second major aspect of BAS energy management is the orderly switching of blocks of load in the event of an emergency. Few buildings have total standby power capability; therefore, in the event of an outage, it becomes necessary to select which loads and functions should be served with the reduced power available. In a building with computer loads or other critical functions, those critical functions would be maintained at the expense of the normal systems. The following is a somewhat typical BAS operation in a building with critical computers and only two generators serving part of the total load.

When high peak electrical loads are anticipated or recorded, an emergency generator can be brought on the line to peak-shave the load. This approach, however, would require an expensive synchronizer and generator cross-current suppressors in order to synchronize emergency generator with utility power, especially where power motor loads are involved. At peak demand times, normally only one elevator would be in service if there are multiple elevators in the building.

When utility power is completely lost, all building systems (except for the critical computers and possibly some computer area lighting supplied by an uninterrupted power system) will shut down for about ten seconds before the generators come on the line and assume the load. Once the generators assume the load, the BAS automatically connects the computers, computer lighting and air conditioning, security systems, and one elevator at a time to generator power. The building lighting load can be reduced in half, either through the possible use of an automatic dimming system or by automatically turning off half of the fluorescent lamps in the lighting fixtures.

If one of the generators should fail during the outage, cutting available emergency power in half, the BAS would automatically shed all lighting (except for scattered emergency fixtures), the main building HVAC system, and the least critical computers and security systems (such as closed circuit television). Normal operation would be maintained for the most critical computer functions and attendant air conditioning, elevator and critical security systems (intrusion and fire detection).

When normal utility power is resumed, the BAS will automatically turn "on" all building systems in order of critical importance, i.e., computers with their attendant lighting and air conditioning, normal lighting and the remainder of the building HVAC, and the remaining elevators. If the entire building load were switched "on" at the same moment, the resulting inrush current could cause a severe voltage drop through the primary and secondary windings of the transformer coils or cause the switchgear overcurrent devices to malfunction and open-circuit.

In this hypothetical building with critical computers, the lighting load would not have to be completely shed all at one time. It could as easily be half-shed in a completely automatic manner, a manner even independent of BAS control. To accomplish this, separate branch feeds are run to each of the ballasts in every 2 × 4 foot 4-lamp, 40-watt RS fluorescent fixture; each lighting panel would be of the split bus type, fed from both normal utility and emergency generator power sources.

Figure 3–2 illustrates this simple, yet effective system. During utility power outages, the lighting panelboard is fed with generator emergency power which has a limited capacity. This emergency power is principally reserved to power the critical computers. (See Figure 3–3.)

During a utility power failure, the contactor in the split bus panelboard drops out half the branch circuits. Turning off 50 percent of the lamps not only reduces the lighting levels by half, but correspondingly reduces the power consumption by 50 percent. (See Figure 3–4.)

Ballasts controlling inboard lamps are connected to a phase wire in the upper split bus section which receives normal utility power and which, during a utility power failure, receives emergency power. (See

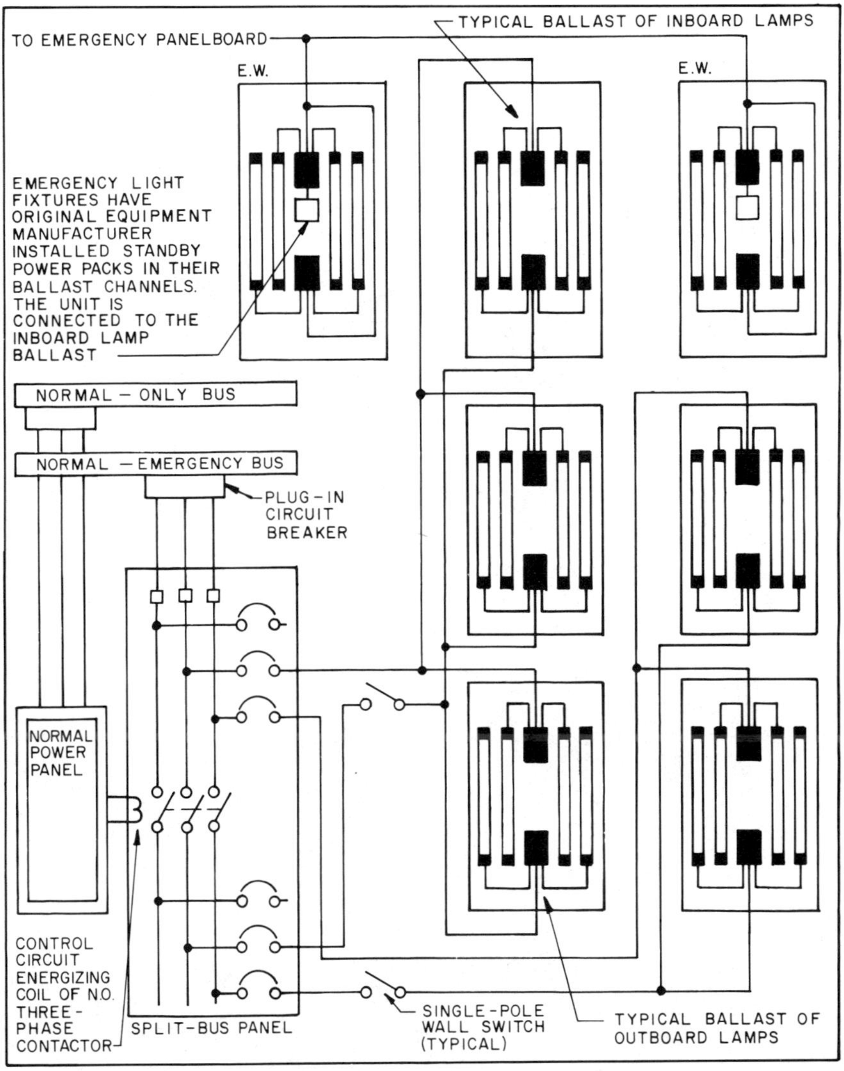

Figure 3–2. Lighting—emergency load shedding detail.

Figure 3–5.) Ballasts controlling outboard lamps are fed from the lower split bus section via a switch leg connected to a normal utility phase wire. This drops out automatically when the contactor senses utility power loss. (See Figure 3–6.) The accompanying photos (Figures 3–3 to 3–6) illustrate the operation of this automated load shedding system.

Figure 3–3. During availability of normal utility power all 4 lamps in a 4 lamp 2 × 4 lighting fixture remain on. This produces 100% illumination levels at corresponding 100% energy use.

Figure 3–4. At the discretion of the computer technician half of the lamps outboard lamps can be switched off. During a utility power outage when only emergency generator power is available these same lamps are automatically extinguished cutting the lighting energy load in half.

Figure 3–5. The outboard lamps are fed power from the lower section of a split bus panelboard and under a utility outage are dropped from service. The inboard lamps are fed from the top section of the split bus panelboard and are fed normal utility power when available but during a power loss are fed only emergency generator power.

Figure 3–6. Under a loss of normal utility power the outboard lamps are extinguished, thus producing 53% of the normal lighting level but at only 50% of the previous energy use.

Certain selected emergency white lighting fixtures should be provided with OEM-installed standby power packs within their ballast channels. There is no power available for emergency lighting for several minutes while the various essential loads are being sequentially picked up during the transition from utility to emergency power.

SECURITY AND SAFETY SYSTEMS

The BAS also has the inherent capacity to monitor and control all life safety, building security and emergency power systems. If a fire alarm or sprinkler system is activated, or smoke detectors fired, the system would automatically send local and remote signals. The system should have the capability to indicate the location of the fire and smoke and be capable of controlling elevators. Prerecorded messages or actual voice commands can be sounded on a public address system. The system may be designed to have the capability to energize fans and dampers to clear smoke and deenergize other fans and dampers to prevent the spread of smoke and fire.

It should also have the capability of monitoring main entrances on closed circuit television and be capable of opening or closing these doors as necessary.

Data processing departments, loading docks or other sensitive areas can be monitored with motion detectors and provided with electronically controlled (card or key) access.

MEETING EMERGENCY POWER NEEDS

Thus far, a building system with critical computer functions has been studied, one with sufficient emergency power available to permit selective management of various building loads. However, where critical functions other than life safety systems are not used, a relatively small backup power source may be sufficient both to serve life safety loads in an emergency and to shave peak demand during periods of normal building operations.

ENERGY MANAGEMENT

On one such project the building was provided with a building automation system in order to supervise building security and monitor the HVAC systems. In addition, the BAS had the inherent capability to manage building loads, thus conserving energy in general and, in particular, reducing loads during periods of peak demand. This in turn reduced utility demand charges to the owner.

The BAS minimized the demand charges by regulating the building energy use within preset guidelines, with the assistance of a computer memory to anticipate periods of peak demand based on the prior building history. Energy use was thus regulated through such techniques as:

- Heating or cooling the building gradually in early morning hours rather than using maximum HVAC input beginning at the start of the working day (6 A.M. to 8 A.M.).
- Automatically reducing lighting levels (in this case through the use of a remote, low-voltage lighting, switch-off system) prior to peak summer air conditioning hours, both to reduce the amount of heat introduced into the space and to offset the increase in AC energy used to remove that heat as well as to counteract the outside heat and reduce the electric energy consumed by the lighting system.
- Minimally reducing building lighting or wintertime heating loads during the peak lunchtime electric cooking period.

BUILDING AUTOMATION SYSTEM (BAS)

This project is a multidiscipline building which performs many complex functions, housing and handling sophisticated equipment and material, and is therefore equipped with a building automation system. In addition, energy conservation and the future trends in the cost and usage of energy by necessity require a planned building energy management system.

A modular computerized building automation system for this type of building provides maximum comfort from the heating and air conditioning systems while minimizing the operating costs.

BUILDING ENERGY MANAGEMENT

This project has building energy requirements whose usage varied hour by hour, day by day, with weather and occupancy influencing cooling, heating and lighting needs. The energy in the building was also used by a small group of computers and other processes.

Building energy management requires continuous monitoring and control of dynamic building functions and of the actions and commands by building operations personnel.

A computerized system such as this one can and does coordinate all automatic and operator related application programs. It also has additional capacities that can be expanded in the future. Its purpose is to schedule application programs, allocate memory resources, interface to input and output peripheral devices, and handle interruptions. It has the inherent capability to establish priorities for all programs and operations. This particular system is capable of controlling and operating programs. The inputs can be compared to preprogrammed target peaks stored in the memory. These peaks are simulated by the computerized simulation of the energy requirements of the building or by the operating history. If the logic forecasts that a new, higher peak will be established during a demand interval, electric loads can be automatically shed to avoid a costly penalty. The loads are shed according to a first loadshed priority, that is, the loads that will be missed the least will be shed first. If this does not reduce demand sufficiently, other loads in the "load cycling" program will then be dropped. If this proves inadequate, the loads are then shed on the basis of predetermined second priority items.

If electric demand still continues to grow, a demand alarm device will sound, and the operator is manually able to shed additional electric loads. Loads are shed on a rolling priority basis; so that the same loads will not necessarily be dropped repeatedly and, in addition, their priorities can be readjusted through a keyboard. The system has the capability for the program to be overridden manually when necessary to leave certain loads on.

The building automation system can also be programmed for remote equipment start/or stop based on time of occupancy and change in outdoor temperature to reduce energy consumption. The program can put the mechanical systems' operations in an automatic mode; however, the operator can override the program.

Lighting zones can be manually tied into the building automation system by being adjacent to the BAS, thus allowing the operator a free choice in the load shedding of the lighting; or the automatic lighting control schedules can be programmed into the system. Occupancy times, weekend and holiday data, janitorial schedules can also be entered into the program, and lights can then be automatically turned on and off, by zone. An operator override feature is also provided. This facility, at present, requires the operator to make all manual load shedding based on the BAS-supplied information. Changes to the program may also be achieved by keyboard entries.

CONCLUSION

This chapter has considered the operation of building automation systems in relation to both energy management and emergency performance. In addition, it has explored related aspects of lighting control and backup power systems which contribute to the smooth functioning of a modern building.

Today, all buildings are becoming machines for living, working and playing—complex machines which must, in an era of energy consciousness, function much more efficiently than they ever were required to in the still recent era of cheap energy. Building automation systems are already an important tool in the achievement of energy objectives, and will, with increasing development and refinement, become more so in the future.

4 INTERDISCIPLINARY ENERGY AND COST SAVINGS

INTRODUCTION

In light of recent technological changes made within the past few years and the scarcity and high cost of fuels it behooves us to seek additional and better methods of energy conservation. Any building that was constructed prior to 1974 was designed during the era of plentiful, abundant and cheap energy. For those reasons, energy wasteful systems should be modified or redesigned in order to save energy as well as money for the owner or renter of a condominium townhouse office or the renter or owner over the life cycle of the office or residence. In many cases, the first costs of renovation and retrofit from these energy saving techniques can be paid back in less than one year. The accrued savings not only pay back the additional cost of renovation and retrofit, both in terms of principal and interest, but save on energy and fuel bills as well.

DESIGN TO LIFE-CYCLE COST ETHIC

The following design parameters concerning the color of asbestos roof shingles (tiles) should be considered before the selection or design of a residence, apartment or condominium office complex. Dark colored roofing material such as black, dark green or brown, have more maintenance problems and shorter usable lifespans. They can also cause higher air conditioning (A.C.) loads, higher energy consumption and thus higher electric bills in the summer. This approach increases the energy expenditure for mining, refining and installation of more finished goods, and also increases future energy use and accompanying maintenance costs because the darker roof material increases the air conditioning loads. Light colored roof shingles not only reflect light and absorb less heat but allow the material to operate cooler during the entire day which reduces attendant A.C. energy consumption and bills. The roof shingles also last longer. A roof of light-colored material may last upwards of 30 years, whereas very dark roof materials may last only 15 or 20 years.

The reason for the difference is that the darker roofing material is subjected to greater, more frequent thermal stressing. The darker the material, the closer it approaches a *black body*. This means that a lighter roof reflects more incident light while a darker (i.e., black) roof absorbs more incident sunlight and converts it into heat (thermal energy). During the night the darker roof radiates more heat energy into space and into the surrounding atmosphere, but during the day it absorbs more radiant energy (sunlight) and converts it into heat. With intermittent cloud cover during the day, each time the sun goes behind the clouds, the roof starts radiating heat into the surrounding area again. When the sun bursts forth, the roof heats up again. This continual heating and cooling from night to day to night is bad enough, but the more rapid, cyclical heating and cooling of the roof during daytime cloudiness or inclement weather does the major damage. Each time the roof heats up it expands, and each time it cools down it contracts, which stresses and fatigues the roofing material leading to its eventual failure.

The darker roof, already more prone to quicker thermal failure through its sensitivity to changes in sunlight intensity, thermally stresses itself even more in going from sunlit areas (thus high heat-gain areas) to shaded areas (high heat-loss areas) caused by an adjacent roof or soffit overhang, producing additional destructive lateral forces. This in turn produces more tearing of tiles and popping out of tile retaining nails, thus increasing repair maintenance costs of the roofing material.

Worse yet, the darker roof not only absorbs more sunlight and thus more heat (sunlight is converted into thermal energy), it also heats up the attic (it elevates the attic temperature by heating up the air in the attic). As a result, on a sunny, 100°F day the attic temperature may be as high as 120°F to 130°F. Although you wish to maintain a cool and relatively comfortable 80°F temperature inside your home, you will have upwards of a 50°F temperature differential between your attic space and the interior air conditioned spaces, despite only a 20°F temperature differential between inside and outside air. As a result of this large temperature differential (ΔT), the added heat is driven back into the office condominium or dwelling by radiation. This in turn increases the time your air conditioner must operate as well as the load on your A.C. system. Thus, there is a proportional increase in the electric-energy used to operate your A.C. system and consequently an increase in your electric energy bill.

The darker roof does not contribute nearly as much heat over the same period in the heating season as it does during the cooling season. Thus, it will not contribute as much beneficial heating in the winter months, to help offset gas or oil heating loads, as it contributes adversely in the summer, when it increases the electric cooling energy use. During the summer solstice the angle of inclination of the earth's axis of rotation is toward the sun. In the northern hemisphere this means that the angle of incidence of the sun's rays is nearly perpendicular to the earth. This allows more sunlight to enter the earth's atmosphere and heat the surrounding earth and air. Since almost all of the incident sunlight is absorbed by the northern hemisphere, it warms up during the summer months. At this time the southern hemisphere is in its winter solstice.

During the winter solstice the angle of the earth's axis of rotation is oblique to and away from the incidence of the sun's rays and therefore more sunlight is reflected off the upper atmosphere. Since less incident sunlight is let through the upper layers of the atmosphere due to its higher angle of incidence, less sunlight is absorbed by the surrounding atmosphere, plants and earth of that hemisphere. As a result, that hemisphere is in its winter solstice and cooling down. Thus, for a given latitude there is more incident sunlight to deleteriously heat a roof in the summer than there is to advantageously heat the roof in winter.

Only extremely northern or southern latitudes might benefit from the heat gain obtained by darker roofs. (See Figures 4–1 and 4–2.) Because of the high grazing angle of incidence of the sunlight at these higher or lower latitudes, those areas remain the coldest and for the longest periods of time. In those areas, the benefits of the heat gain

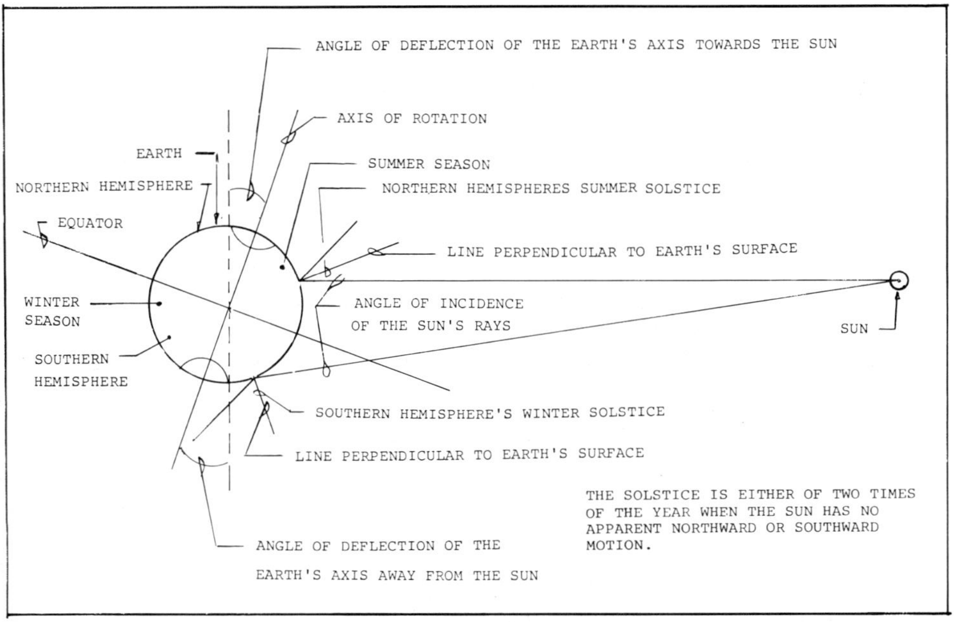

Figure 4–1. Sun's spatial relationship to the earth.

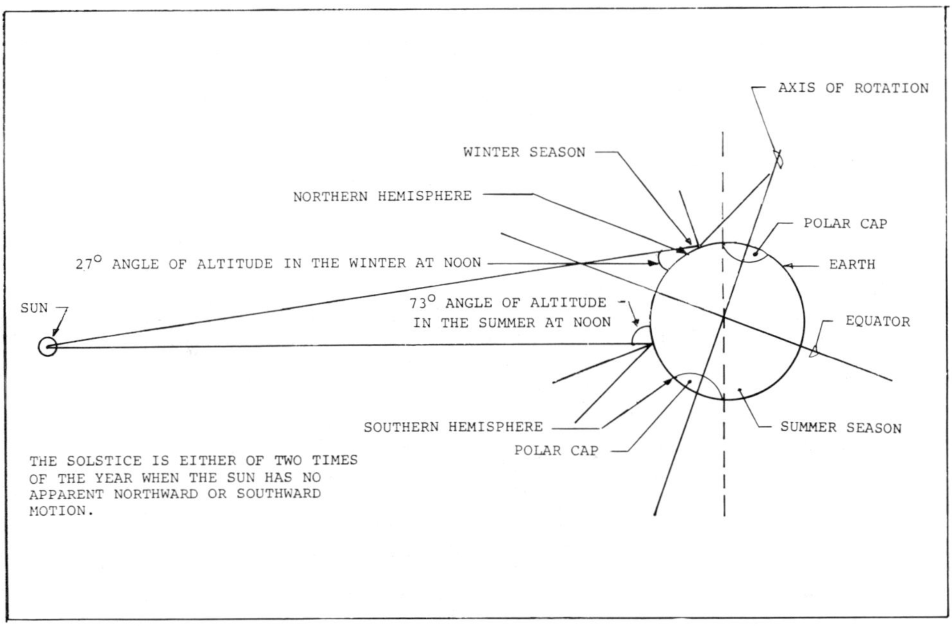

Figure 4–2. Sun's spatial relationship to the earth.

may help to offset the cost of winter heating at no appreciable deleterious effect to the roof. In fact during winter months more heat may be radiated by the darker roof at night or during cloudy overcast days than is collected during the entire portion of a sunny day. With proper and sufficient attic insulation most of the heat loss should be minimized so that for the same period of time the heat gain will exceed the heat loss. Since during the winter months the days are far shorter, there will be far more time when the roof radiates heat out into the surrounding space. During the winter months there will be nearly twice as much time when there will be darkness due to night and overcast days during which time the roof will be radiating heat. The further north in northern latitudes or south in southern latitudes, the shorter the period of sunlight during winter months. At the much higher latitudes you approach an area that remains in total darkness for months at a time. At such locations a darker roof would certainly offer no heat gain advantages over a lighter roof. There is insufficient heat build-up, transient or thermal stressing to seriously accelerate the aging and deterioration of a darker roof at these latitudes. In addition, there is really no need or requirement for air conditioning in these extremely higher latitudes.

As a rule, summer cooling is usually more energy intensive than winter heating. As a result, the cost for electric energy use for summertime air conditioning is higher that the cost of winter heating with gas or oil. Although the duration in central North America for the additional summer heating loads requiring additional cooling is nearly equal to the duration of potential winter heating benefits, the additional energy cost disadvantages of summer cooling far outweigh the winter heating advantages and cost savings.

One of the contributing factors to this is that gas and oil are also used to generate electricity which in turn operates electric air conditioning equipment. As a result of the heat energy conversion of gas or oil into mechanical energy in turn converted to electric energy, there is a great deal of loss in efficiency. This greater loss in energy conversion efficiency in going from gas or oil to electricity makes it more energy inefficient, energy consumptive and costly. There are additional line losses in transmission as well in transforming it from one voltage to another. When resource utilization factors are used in the analysis it can be shown that of the 3.04 Btu's of electric energy produced from oil only 1 Btu of electric heat energy is produced in the home. It takes a great deal more Btu of oil-produced electricity to produce 1 Btu of household air conditioning. As a result, the benefits of possible potential defrayed winter heating energy use and cost do not offset the deficiencies of added energy use and cost for summer cooling.

Regardless of whether you have light or dark roofing material, your A.C. load, its attendant electric energy usage and its companion energy cost can be reduced by attic exhaust fans. Most residential homes, professional office rental apartments and office condominiums are built with an A-frame gabled roof with screened openings in the eaves and framed louvered vents at the top front or apex of the A-frame roof. Cooler, denser, heavier air which is lower to the ground is drawn up through the eaves by convection. This cooler air is heated, and as its temperature increases it expands, thus becoming lighter. It is thus exhausted through the louvered A-frame vent. (See Figure 4-3.) This is the principle of convection. But nature can be improved upon. By forcefully venting the attic space with fans to increase the circulation, the temperature within the attic space can be drastically reduced. (See Figure 4-4.)

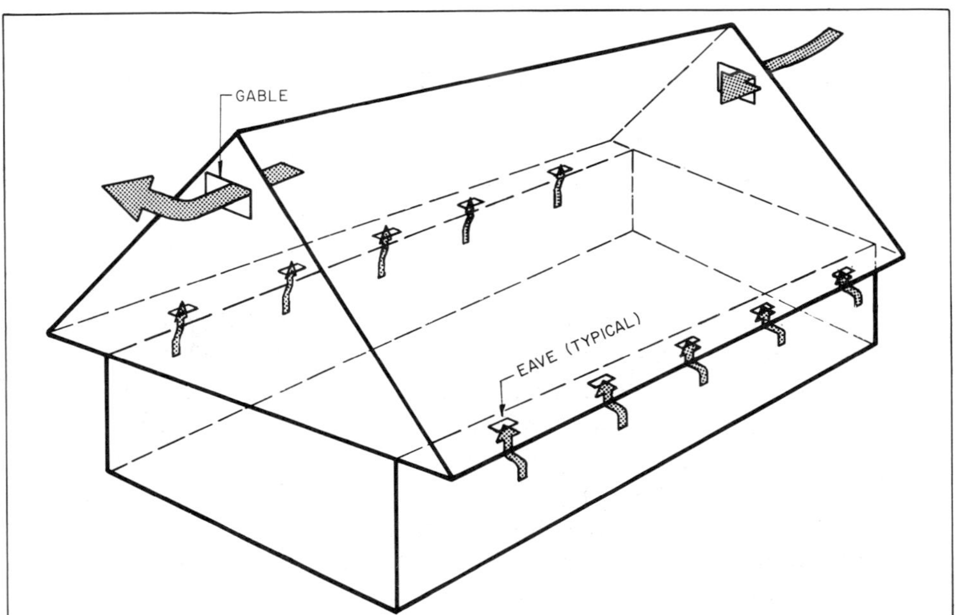

Figure 4-3. Natural convection cooling of attic space.

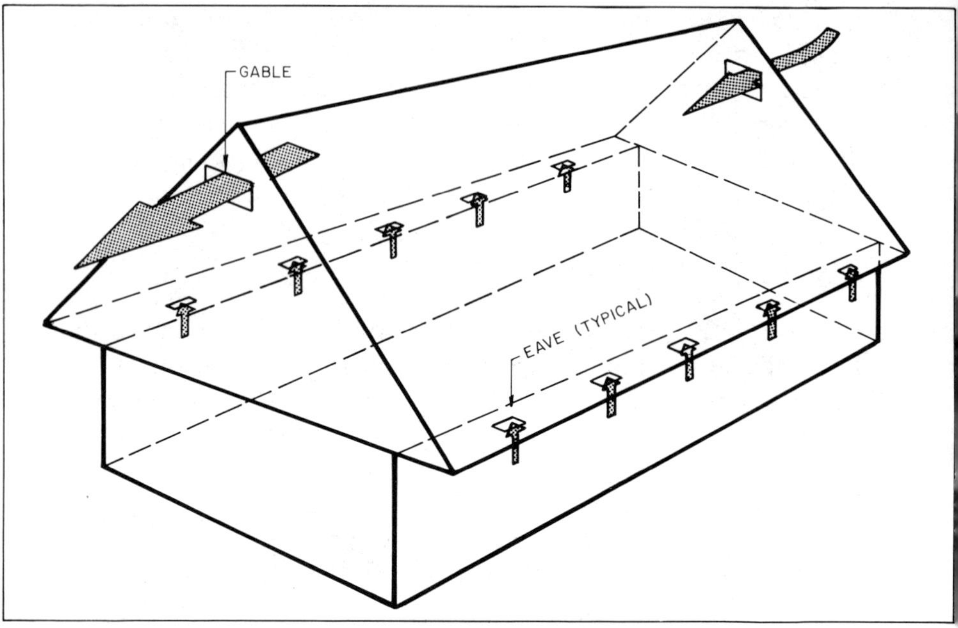

Figure 4–4. Forced fan induction method for cooling of the attic space.

A resident or office worker desiring an 80°F inside temperature when the outside ambient air temperature is 100°F will have the following condition: 100°F air is quickly brought into the attic space and exhausted by fans, thereby holding the attic temperature in the 100°F to 110°F range. Thus, only a 20°F or 30°F temperature differential is experienced. By using attic fans a decrease in temperature differential of from 50°F down to 20°F or so drastically reduces the air conditioning load and the electric energy consumed to operate it. This in turn reduces your summertime electric bills. The fans are actuated by a thermostat when the temperature in the attic exceeds a preset temperature limit. Although these fans may run continuously all day on hot, bright, sunny days this energy load is far less in proportion to that which would have been consumed by the A.C. equipment had the attic exhaust fans not been used.

If you own a split foyer, two-story or rambler home or office condominium having only one common A-frame roof, usually one $\frac{1}{10}$ to $\frac{1}{5}$ hp exhaust fan is needed. A split level house or condominium office having two levels and thus two separate roofs would require two exhaust fans. For a condominium office or residence having a 3-ton (36,000 Btu) air conditioner, and a central furnace with a $\frac{1}{3}$ to $\frac{1}{2}$ hp air handler to blow air over the A-frame cooling coils, the total A.C. electric load will be about 6.5 kW. This is because the outside patio or pad-mounted condenser or direct expansion unit will have at least a 3 to 4 hp compressor motor, a $\frac{1}{2}$ to $\frac{3}{4}$ hp condenser fan motor and one or two $\frac{1}{3}$ to $\frac{1}{2}$ hp condenser supply and/or return pumps. This condenser load, in conjunction with the air handler motor, will thus use about 6.5 kW, 12 hours a day, 30 days a month, for about 2367 kilowatthours (kWh) of energy consumption per month. At 6¢ per kWh, this comes to no less than $142 every month for air conditioning to provide creature comfort.

In a split-level condominium office or residence with one or two $\frac{1}{10}$ hp attic fans, the energy use for 12 hours per day, 30 days a month will be 228 kWh, for a cost of $13.70. The same central A.C. load will now be reduced to anywhere from $\frac{1}{3}$ to $\frac{1}{2}$ the previous level, providing that the condominium office or house is reasonably insulated, caulked, and has storm windows and doors. Thus, in lieu of spending $142 per month on air conditioning, your bill may be reduced to anywhere from $61 to $85 monthly, at monthly savings ranging from $57 to $81.

In most instances there will be many days in which the attic fans will have sufficiently reduced the radiated heat load injected into the condominium office or residential space in conjunction with lower temperature days (below 90°F) and lower humidity days whereby local oscillating office or household fans will suffice.

Another approach to be considered, used either by itself or in conjunction with light colored roofs or attic fans, is shading. Planting shrubbery adjacent to the building and planting what will eventually become large shade trees in close proximity to the house or office not only reduces summer air conditioning loads but also helps to reduce winter heating bills.

CONCLUSION

Each of these approaches is cost effective by being efficient and energy conservative by itself, and each concept and design approach imparts energy efficiency to the various other disciplines. With the advent of the Arab oil embargo in 1973 and the Iranian crisis in 1980, oil prices and energy costs in general have skyrocketed out of sight. As a result of the high cost of electricity, it behooves us to find ways to effectively reduce electric energy usage without sacrificing creature comfort, goods and services, or working an undue hardship on office workers or homeowners. In a very brief way the approaches outlined indicate alternate approaches to today's dwindling fuel supplies and suggest new ways to deal with the energy "shortage." We have addressed energy conservation through good design practices without engendering great personal sacrifice on the part of the American people.

5
WATER-CONSERVING TECHNIQUES THAT ALSO CONSERVE ENERGY AND MONEY

INTRODUCTION

The following narrative and analysis will consider water-saving plumbing technologies and design techniques that contribute to minimizing energy consumption and building operating costs. We are seeing rapid technological changes in the construction industry in this field. We will continue to see rapid changes in the residential, commercial and industrial fields in the years ahead. The industry is only just now on the cutting edge of advances in the field of innovative technological changes that affect our life styles by conserving our precious resources which are in limited supply.

WATER-SAVING, ENERGY-SAVER TECHNIQUES

One area of energy conservation that should have more importance attached to it and is not seriously considered is water conservation through reducing consumption. There are many water-saving techniques and instruments on the market that reduce the quantity of water consumption while not hindering or reducing its usage and usefulness.

Many conventional, older-design water closets (tank-type water bowl) store and use 3.5 gallons of water per flush. The more conventional, but older designs of flush type value toilets use from 3.5 to 4 gallons per flush. The same applies to older urinals in that they require from 2.5 to 3 gallons per flush.

The newer, tank-type water-use limiters are able to get by on 2 gallons or less per flush, while the latest flush-type value toilets utilize as little as 2 quarts per flush. The latest in water-flow-control, value-type, urinals only require $\frac{3}{4}$ to 1 gallon per flush. Sink water faucets can now come supplied with self-closing faucets, stop valves and faucet water-control valves that limit water consumed either by reducing the flow, restricting the time of flow for each actuation or both. The latest in shower heads utilizes an aerator in conjunction with a restricter valve. This allows you to get a good washing wet-down action while utilizing only 40% or less water. In addition, it has a shut-off drip feature at the shower head. While soaping up in the shower the shower is turned off at the shower head in order to conserve water when it is not needed. The drip feature allows the water to remain at the same temperature so that upon being turned back on, you are not turned into a block of ice due to the water having set for a while and becoming relatively very cold.

Obviously, substantially reducing the flow of water per flush in a toilet or urinal not designed to accommodate the lower flow rate adversely affects its performance. An example of just such an approach is placing bricks or other heavy, water-displacement objects in the tank of a tank-type flush toilet. It reduces the quantity of water for each flush but may not remove all liquid wastes and probably will not remove all the solid wastes. In some cases two or more flushes are required, thereby actually increasing the amount of water consumed by such ‘‘ingenious’’ techniques.

With equipment designed to reduce the quantity of water used while properly performing its function, the same amount of water-related activity, i.e., washing hands and using other toilet facilities, can be performed while conserving a very precious and expensive commodity —water. In large office buildings or residences, a hot water recirculation pump can also conserve on water usage. By having hot water available at the tap instantly, an individual does not need to run the hot water tap for extended periods of time in order to get instant hot water by emptying the line of the cooled water.

It can be argued that in low rise (12 floors or less) office and apartment/condominium buildings, lift-pump energy savings are not directly accrued by the owner through energy savings within his building and equipment. In most locations adequate pressure exists from water-tower tanks to supply the proper hydraulic pressure. The public water authority has to expend electric energy and its attendant cost for lift pumps to elevate and lift the water to these tanks during off-hour electric use and off-peak water use. Their electric energy costs are then passed on by the water authority to the owner or end user of this product, water. In such cities as New York and Chicago with their high rise office complexes, water tanks for storage and pressure can be viewed on almost every roof top. In some cases totally enclosed, closed-cycle, pneumatic pressure tanks are incorporated. In both cases it takes the use of a lift-pump electric motor to elevate that quantity of water to that greater height in excess of the water-level height that can be obtained under normal domestic pipe line pressure. As an alternative some high-rise facilities may incorporate variable-speed, booster pressure pumps which are normally driven by electric induction motors as opposed to gasoline or diesel engine driven units.

The many advantages of the various water-conserving approaches are reductions in first cost, water and companion sewage treatment costs, energy and life-cycle costs.

Lift-pump motor electric energy is conserved; less energy is consumed in processing river water into clean potable water; less energy is also consumed in processing sewage especially when most of the effluent may be made up of otherwise wasted "clean water." Gas, oil and electric energy is saved in either domestic or central hot water heating systems since less hot water will be wasted and less energy dissipated using a hot-water recirculation pump. Not to be overlooked is the money and energy savings in some areas of the country by not having to construct more or larger sewage treatment plants, especially where there are bans on new sewer connections. More toilets, faucet and sink connections can be added to the building if all the existing and newer sewer connections within the expanded facility are water-saving devices whose total volume of water consumption and discharge will remain the same or be less than with the fewer original non-water-saving plumbing fixtures. By having new residences, condominiums and apartments utilize these water-saving and conserving devices, existing sewage treatment plants don't have to be expanded or modernized. These savings are then passed on to the homeowner either in lower water and sewerage bills and/or lower county taxes that would be assessed for sewerage treatment plant expansion.

Consider the following analysis of lift-pump energy use in a modest size office building that can be found in Washington, D.C. Then multiply that use by the added size and height of buildings in New York, Chicago, etc. to realize the ramifications and impact factors that can be involved. As has previously been analyzed, consider the building height restriction in Washington, D.C. which precludes visually obscuring federal structures and monuments such as the Capitol and the Washington Monument. For the Pennsylvania Avenue Development, district C–5, 130 feet is the total height restriction. This restriction will allow a 10-floor structure with a floor-to-floor height of 13 feet, or 12 floors at 10′–10″ floor to floor. We will assume the latter and in addition will assume that one water closet or urinal will be flushed per floor every 2 minutes. Further it is assumed for simplicity that the water savings between either a conventional water closet or urinal and a comparable water-saver counterpart will be 3 gallons. Each gallon of water weighs 8.34 pounds and there are 7.5 gallons per cubic foot of water or, more specifically, every cubic foot of water weighs 62.55 pounds.

We will figure that the first floor of this hypothetical office building or complex is at ground level and therefore no lift pump energy is required to supply water to these various toilet fixtures. For the sake of simplicity we will stipulate that there is an 11′ rise to the next level or second floor in lieu of the exact 10′–10″ just derived and a 22′ rise to the third floor etc. Thus there is a total accumulative rise or *head* of 726 feet based on the rise to each consecutive floor level of 11, 22, 33, 44, 55, 66, 77, 88, 99, 110, and 121 feet. Since there are only eleven floors that are above ground, we are thus only concerned with those toilet fixtures being supplied on those floors. Therefore there are eleven water-saver water closets or urinals being flushed every 2 minutes, but each is using 3 gallons less per flush. The total lift energy saved is computed by multiplying 11 toilet fixtures times 3 gallons saved per flush times 8.34 pounds per gallon times 726 total feet of rise (head) divided by 2 minutes = 33 gallons (8.34 lbs.) 726 feet divided by 2 minutes = 199,809.72 foot-pounds/2 minutes = 99,904.86 foot-pounds/minute.

$$\frac{99{,}904.86 \text{ foot-pounds/minutes}}{} \div \frac{33{,}000 \text{ foot-pounds/minute}}{\text{hp}}$$

$$= 3.027 \text{ hp (say 3 hp)}$$

Each hp represents 746 watts of electric energy. If the total combination of electric squirrel cage motor efficiency and pump efficiency is 0.5 then there will be 4476 watts of input energy (3 hp × 746 watts/hp divided by 0.5 EFF) required. If this facility is in continuous operation 12 hours a day, 5 days a week and 52 weeks a year, then the electic motor will run 3,120 hours for the year.

At 6¢ per kilowatthour (kwh) this represents $837.91 saved every year just in lift pump electric energy alone (4476 watts divided by 1000/watts/kW) × 3120 h/yr × $0.06/kWh = $837.91). There will be the cost savings for the additional water conserved as well as the savings accrued for the reduced companion sewage treatment costs.

CONCLUSION

Many equipment manufacturers of plumbing fixtures for both new and rehabilitated work on the market today incorporate the technologies for better resource utilization and conservation as well as energy conservation. There is a vast proliferation of new technological marvels for new designs and installations as well as for retrofitting existing facilities that can conserve our resources while reducing energy costs. Some are more cost effective than others and have a greater validity to the claim of making a worthwhile contribution to our technology and our society as a whole.

We are indeed witnessing a startling revolution in ideas and system components. An even greater proliferation of hardware is at the disposal of plumbing designers and engineers, contractors, maintenance personnel and owners. If properly utilized, they should benefit man in his war on waste and the ever-present challenge of energy conservation.

Specifically the items outlined apply equally well to retrofitted structures as to new construction. In many cases, the first cost of renovation or retrofit can be paid back in less than one year from these resource conserving and energy saving techniques. For retrofit projects the cost of renovation not only saves on water, sewage and electric bills, but the accrued savings pay back the additional cost of the renovation in terms of both principal and interest.

6 ON THE FRONTIERS OF LIGHTING TECHNOLOGY

INTRODUCTION

We have already seen rapid technological changes in the electric lighting industry in the past 10 to 15 years. We will continue to see startling changes in this field in the years ahead. The industry is only just now on the cutting edge of advances in the field of technological change. We are indeed witnessing a startling revolution in ideas and system components. An even greater proliferation of hardware is at the disposal of the lighting designers and engineers, contractors, maintenance personnel and owners. If properly utilized, they should benefit man in his war on waste and in the ever-present challenge of energy conservation.

THE LAST HALF CENTURY IN LIGHTING TECHNOLOGY

In the past 50 years many advances have been made in lighting technology. Notably, General Electric introduced the first commercially available mercury vapor lamp in 1934. GE also did some experimental work on high pressure sodium lamps in the late 1940s. Low pressure sodium (LPS) lamps were first developed in this country in the late 1920s (approximately 1928); however, they have been developed and used much more extensively in Europe for the past 50 years. The most basic modern light source used extensively in this country and throughout the world is the fluorescent lamp, made commercially available by GE in 1938.

In addition to the above electric light sources, the first and most basic one has existed for just over a century, since Thomas Alva Edison first tested his carbon filament lamp on October 19, 1879. The basic incandescent lamp, depending on size, shape and function, has an efficacy of from 15 to 18 lumens per watt and a relatively short life span of from several hundred hours to several thousand. A relatively recent outgrowth of this lamp, developed around the early 1960s, is the tungsten halogen (quartz iodine or quartz halogen) lamp having a slightly higher efficacy and purported longer life, which does not seem to materialize under field conditions. These tungsten halogen lamps are a great deal more expensive but of extremely compact size even though linear in shape, they are ideally suited as point sources where a high degree of control is required by a reflector system. Their shorter than advertised lifespan may have been due in part to poorly designed luminaires, heat sink bodies and heat dissipation fins.

There will always be some areas where aesthetic requirements such as color rendition may make the incandescent light source preferable. In such cases where they are the preferred light source, consideration should be given to using either tungsten halogen or ellipsoidal reflector (ER) lamps. Both types are more efficient, in terms of light delivered per watt of power consumed, than conventional incandescent lamps. Due to the ER lamp's unique reflector shape, it focuses the light beam at the aperture of a baffled, deep-recessed downlight. The ER lamp focuses its beam two inches in front of the lamp where it then spreads out and thus allows very little light to be trapped by the baffles of the downlight lighting fixture, whereas a conventional reflector flood lamp projects a wide conic beam of light of which 60% can be trapped and wasted in the baffles of a highhat downlight. (See Figure 6–1.) As a result, lighting efficiency is doubled with ER lamps when compared to the use of conventional reflector flood lamps. In many instances an ER lamp of half the wattage of a standard reflector lamp can be used to produce the same illumination level at the floor or task. (See Figure 6–2.) In many cases the energy savings over the 2000-hour rated life of the lamp may be upwards of $7.50 or more per lamp and fixture combination. (See Figure 6–3.)

Within the last decade the following advances have been made in lamp technology. The MV lamp family is now considered a basic-

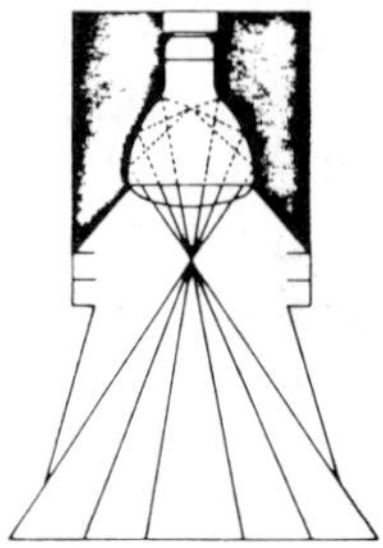

THE ER BEAM FOCUSES TWO INCHES AHEAD OF THE LAMP, THEN SPREADS OUT. VERY LITTLE LIGHT IS TRAPPED BY THE FIXTURE BAFFLES.

A REFLECTOR FLOOD LAMP PROJECTS A DIVERGENT CONE OF LIGHT. AS MUCH AS 60% IS TRAPPED BY FIXTURE BAFFLES.

ER LAMP SPECIFICATIONS:

ORDERING CODE	VOLTS	WATTS	AVERAGE RATED LIFE	MAX. OVERALL LENGTH
50ER30	120	50	2,000	6-3/8"
50ER30/PK	120	50	2,000	6-3/8"
75ER30	120 OR 130	75	2,000	6-3/8"
120ER40	120 OR 130	120	2,000	7-3/8"

COURTESY OF GENERAL ELECTRIC

Figure 6–1. ER elliptical reflector lamps.

ally obsolete light source but has been developed to have efficacies anywhere from 34 to 61 lumens per watt, depending on wattage size. As a whole, the clear lamps are small light sources and almost the entire MV lamp family enjoys a lifespan of 24,000 plus hours. (See Figure 6–4.) The MH lamps are small light sources that became avail-

COMPARE LIGHT PATTERNS BELOW:

THE BEAM SPREAD OF ER LAMPS IS SIMILAR TO THAT OF PAR-38 FLOOD LAMPS - AND NARROWER THAN R-30 AND R-40 FLOODS.

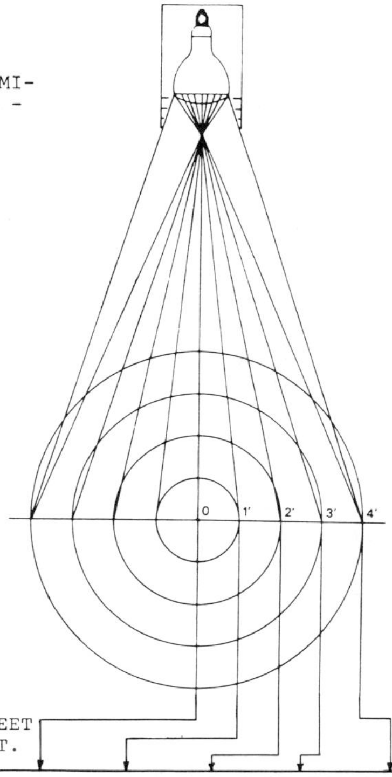

* THE LIGHT PATTERNS SHOWN ABOVE ARE FOR A SINGLE FIXTURE; IN MULTI-LAMP INSTALLATIONS, LIGHT FROM ANY ADJOINING FIXTURES MAY CONTRIBUTE SUBSTANTIALLY TO FOOTCANDLES NEAR THE EDGE OF THE PATTERNS.

** IN DEEP FIXTURES WITH WHITE INTERIOR FINISH. IF FINISH IS BLACK, LIGHT LEVELS MAY BE LESS THAN HALF OF THE VALUE SHOWN.

COURTESY OF GENERAL ELECTRIC

APPROXIAMTE FOOTCANDLE LEVELS 10 FEET BELOW TYPICAL DEEP BAFFLE DOWNLIGHT.

LAMP TYPE	WATTS SAVED	AVERAGE INITITIAL FOOTCANDLES* (at indicated distance from center)				
75R30/FL	25	4	4	3	3	2
50ER30		12	11	9	6	3
75R30/PK	25	9	8	5	3	2
50ER30/PK		7	6	5	3	2
100A**	50	4	4	3	3	2
50ER30		12	11	9	6	3
150R/FL	75	10	10	8	6	4
75ER30		16	14	11	7	4
150R/FL	30	10	10	8	6	4
120ER40		24	23	18	10	5
300R/FL	180	18	17	15	11	7
120ER40		24	23	18	10	5

Figure 6–2. Conventional and ER lamp lighting levels and distribution.

able in 1965 and as a lamp group or family enjoys a 10,000 to 20,000 hour life (with the notable exception of the standard 175-watt, 1000-watt and 1500-watt size at 7500, 12000 and 3000 hours, respectively.) General Electric's 175-watt MH has a life expectancy of 10,000 hrs. In recent years these lamps have enjoyed phenomenal increases in

COMPARE AND SAVE (Courtesy of General Electric)

IF YOU USE:	SWITCH TO:	LIGHT LEVELS (see below)	ENERGY $ SAVINGS PER LAMP*
75R30FL	50ER30	SAME OR MORE	$2.50
75R30/PK	50ER30/PK	NEARLY THE SAME	$2.50
100A**	50ER30	DOUBLE OR MORE	$5.00
150R/FL	75ER30	SAME OR MORE	$7.50
150R/FL	120ER40	DOUBLE OR MORE	$3.00
300R/FL	120ER40	SAME OR MORE	$18.00

* AT 5¢ PER KWH, FOR 2,000 HOURS

** IN DEEP FIXTURES WITH WHITE INTERIOR FINISH. IF FINISH IS BLACK, LIGHT LEVELS MAY BE FOUR OR FIVE TIMES HIGHER WITH A 50ER30.

Figure 6–3. Energy savings using ER lamps or replacements for conventional lamps.

efficacy as new metal additives, arc tube configuration and electronic circuitry have evolved. Ironically, it was the metal additives to the mercury that evolved them from the original MV lamp to the newer MH lamp. They now have efficacies of from 80 to 125 lumens per watt in the so called super metal halide evolution of the lamp. (See Figures 6–4, 6–5 and 6–6.)

In the horizontal burning SMH lamp by Sylvania the central portion of the arc tube is bent upward so that the hot arc's natural tendency to convect upward will not cause the hot arc plasma to make intimate contact with the tube and thus lose its heat energy and correspondingly lose its emitted light output. GE's 400-watt horizontal SMH lamp utilizes a vertically positioned arc tube. Luminaires that utilize the horizontal burning SMH or banana lamp or the GE SMH lamp must have a position-orienting, mogul-base lamp socket so that the lamp will always be correctly positioned in order to benefit from this advance.

The latest breakthrough in SMH lamps by G.E. is the 175-watt Multi-Vapor-II lamp rated at 16,600 initial lumens and 10,000 hours

LAMP TYPE		WATTS (Nominal)	LAMP ORDERING NUMBER		INITIAL LUMENS	LUMENS PER WATT	RATED LIFE
METAL HALIDE	CLEAR HORIZONTAL	1500		M1500/BU-HOR	150,000	100	1,500
		1000	G.E.	MV1000/U	107,800	107.8	10,000
		1000		M1000/BU-HOR	95,000	95	10,000
		400	SYL.	MS400/HOR	40,000	100	•20,000
		400		M400/BU-HOR	•34,000	•85	•20,000
		250		M250/BU-HOR	19,500	78	•10,000
		175	SYL.	MS175/HOR	15,000	85.7	7,500
	CLEAR VERTICAL	1000	SYL.	MS1000/BU	125,000	125	10,000
		1000	G.E.	MV1000/BU/1	115,000	115	10,000
		1000	West.	MH1000/BU	100,000	100	10,000
		400		M400/BU	34,000	85	15,000
HIGH PRESSURE SODIUM		1000		LU1000	140,000	140	24,000
		400		LU400	50,000	125	24,000
		310	G.E.	LU310	37,000	119.4	24,000
		250	G.E.	LU250/BU/S	30,000	120	24,000
		250		LU250	25,500	102	24,000
		150		LU150/55	16,000	106.7	24,000
		150		LU150/100	16,000	106.7	24,000
		100		LU100	9,500	95	•24,000
		70		LU70	5,800	82.9	•24,000
DELUXE WHITE MERCURY		1000		H1000DX36-15	60,000	60	24,000
		400		H400DX33-1	21,750	54.4	24,000
		250		H250DX37-5	12,300	49.2	24,000
		175		H175DX39-22	8,180	46.7	24,000
		100		H100DX38-4	4,200	42.0	24,000
•MET HAL. CLR. VERT		400	SYL.	MS400	40,000	100	15,000

UPDATING FOOTCANDLE VALUES

$$\frac{\text{New Lamp Lumen Output}}{\text{Old Lamp Lumen Output}} = \frac{\text{Updating}}{\text{Multiplier}} \quad \text{Example: 400 W Metal Halide} \quad \frac{40000}{32000} = 1.25 \text{ Factor}$$

Old Footcandle Value x Updating Factor = New Footcandle Value

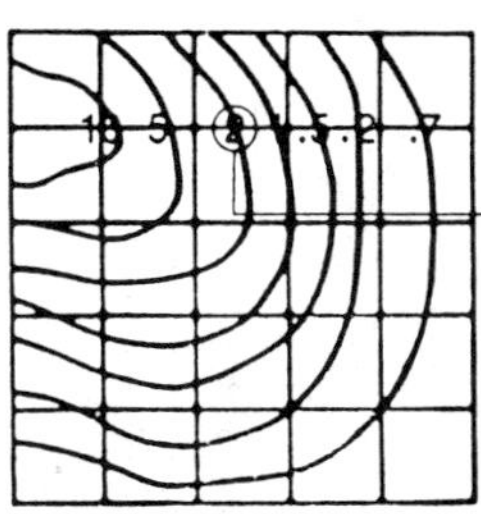

Example:

2FC x 1.25 = 2.5 Initial Footcandles (Based on M400 rated at 40000 lumens).

EXISTING FOOTCANDLES
(Based on M400
Rated at 32000 Lumens).

Lamp data, © 1977, Gardco Manufacturing Inc. Author's amendations are marked by *.

Figure 6–4. Lamp data.

of life. This represents an efficacy of 94.8 lumens per watt versus a standard 175-watt MH lamp rated for only 14,000 initial lumens at an efficacy of only 80 lumens per watt. The new Multi-Vapor-II lamp has the same life rating as the conventional G.E. 175-watt MH lamp. In order to achieve this higher efficacy an advanced electronic integral

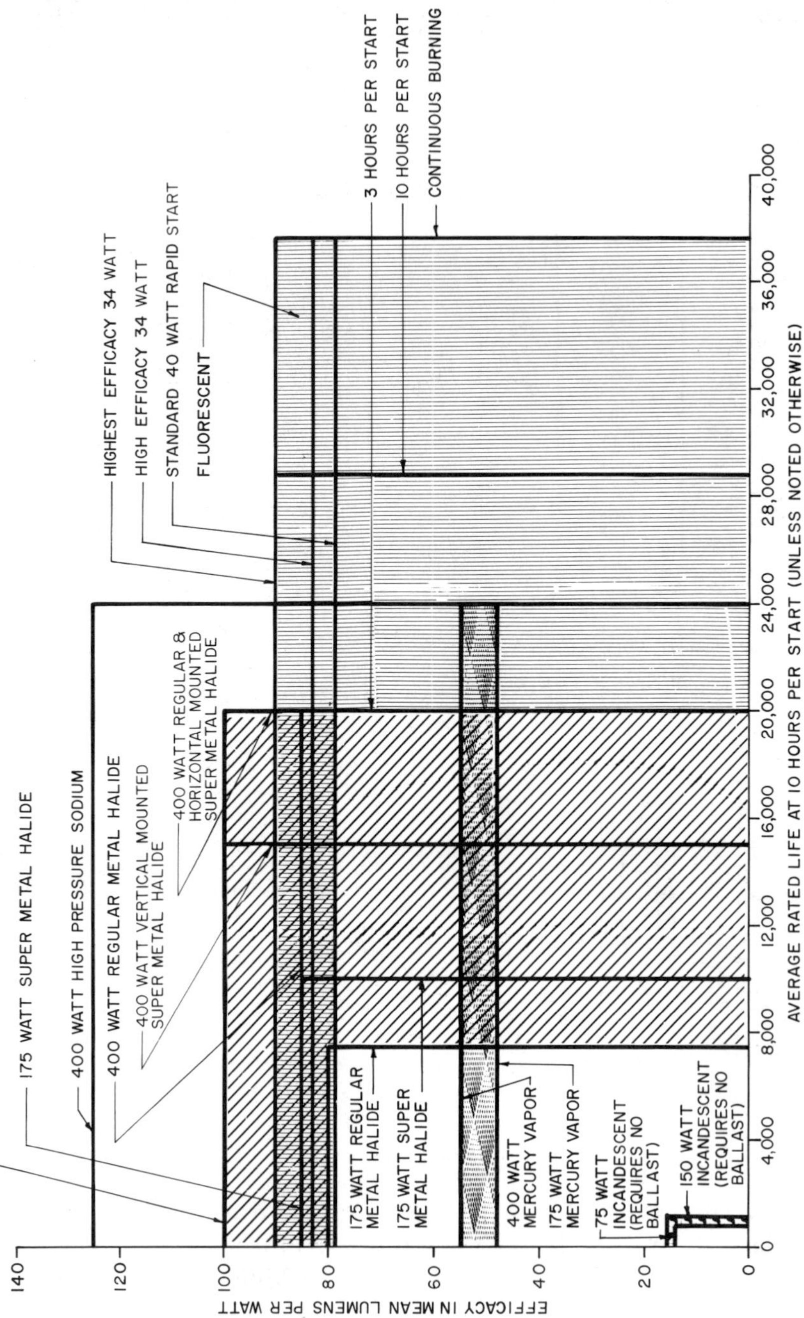

Figure 6–5. Comparison of total light delivered per watt throughout life (without ballast loss).

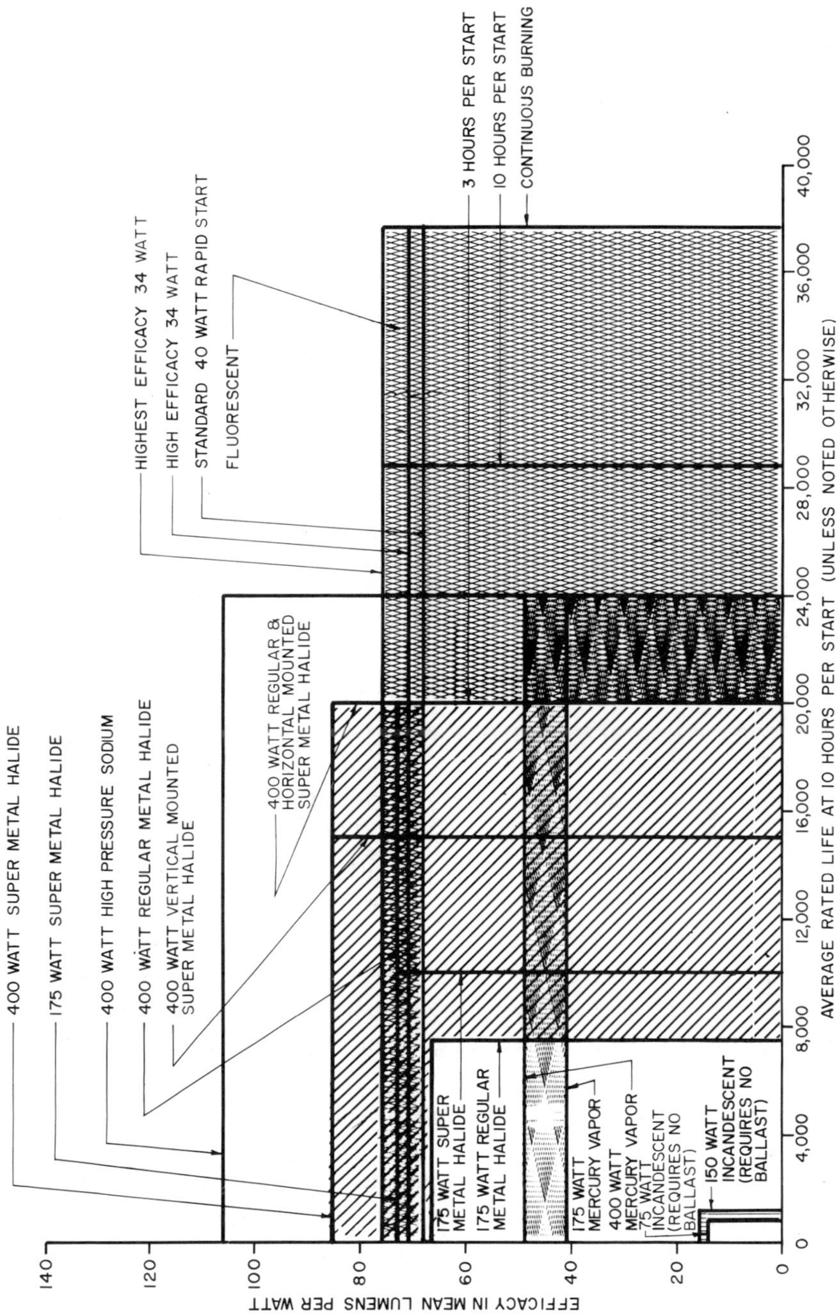

Figure 6–6. Comparison of total light delivered per watt throughout life (with ballast loss).

pulse circuit has been built into it which starts the lamp. The wire probe adjacent to the arc tube is part of the starting-aid circuit to initially ionize the lamp. This also permits this newer lamp to be operated on a conventional 175-watt MH ballast. The color rendering index (CRI) has been improved to produce a warmer pleasing color.

GTE recently announced a family of Sylvania Super Saver Metalarc 3K lamps. These HID lamps are particularly recommended for merchandising applications where both color and energy savings are important considerations. Available in 175-, 250- and 400-watts the lamps produce a "warm" 3200°K color temperature at efficacies ranging from 80 lumens per watt for the 175-watt to 92.5 lumens per watt for the 400-watt. The warmer color is accomplished by a phosphor applied to the inside of the bulb which converts the ultraviolet and blue energy produced by the metal halide arc tube to longer wavelengths concentrated in the red region of the light spectrum.

The horizontal burning SMH lamp by GTE Sylvania comes in the 175-, 250- and 400-watt sizes. A 400- and a 1,000-watt, base-up or base-down, vertical-burning SMH lamp by Sylvania and a 1,000-watt vertical burning SMH lamp by General Electric have their arc tubes narrower at the top and bottom, with a widened middle section. This is to allow the hot rising gases a chance to expand as they rise; and as they become cooler at the top of the arc they contract and once again require less space. As the cooler arc plasma gasses descend, they are able to smoothly flow on the outer layer of the hot rising gasses but are adjacent to the cooler arc tube envelope. This way we get a smoother, nonturbulent, laminar flow of gasses without appreciable eddy currents. In this way the cooler descending arc gasses do not interfere with the hotter rising arc gasses. By reducing the interference, the hot gasses are not forced to impart their heat to either the cooler descending gasses nor are they driven up against the cooler arc tube where they would be forced to surrender even more heat and loss of emitted light.

HPS lamps are also small light sources with extremely long lives and higher efficacies. The HPS lamps have extremely high efficacies of from 82 to 140 lumens per watt depending on wattage size with the notable exception of the smaller lamp wattage sizes such as the 35-watt HPS lamp at only 64 lumens per watt. (Figs. 6-4, 6-5, 6-6.) This lamp with continued development may come close to its theoretical maximum efficacy of 385 lumens per watt. Until recently the HPS lamps produced a dichromatic yellow-orange spectral power distribution (SPD). This golden-hued light rendered colors and complexions poorly and thus was relatively unsuited to interior commercial lighting use. Today its SPD is greatly enhanced, and this, coupled with the proper selection of colors and fabrics color-coordinated and selected under this light source, has done much to overcome its disadvantages.

Another tool in the technological bag of tricks is to illuminate the space to 100 fc or more which in conjunction with the proper selection of color and fabrics physiologically saturates the optical sensing system and cause up to psychologically perceive the color as white. From HPS study and research, Westinghouse has developed an ultra-high pressure HPS lamp as an outgrowth of its new arc tube pressure seals. As a result, a high CRI HPS lamp of much lower efficacy (but still comparable to MH and SMH lamps) has been developed. At present 150-, 200- and 250-watt versions of the high CRI HPS lamp are now available with efficacies of from 91 and 95 to 100 lumens per watt, respectively. Their rated life span is 7500 hours, but it is felt that with improvements in the arc tube seal and extended testing their projected life will be extended to 15,000 hours, perhaps even 24,000 hours. These new high CRI HPS lamps are now widely available, making HPS a viable selection for interior lighting. In fact, high CRI HPS and SMH lamps promise to be the preeminent lighting sources of the next decade.

The ballast for starting and controlling a high CRI HPS lamp is the same as used for conventional HPS lamps. In addition, Jefferson Electric, a division of Litton Industries, has recently (within the past three years) developed a new *Safety Starter* ballast that increases the life of a conventional HPS lamp system. It incorporates an automatic 10-second "cut-off" of start up of the high-voltage pulse. In addition, the ballast stays "off" until the lighting circuit is reactivated. This feature eliminates unnecessary high voltage stress on the complete system, such as the lamp, socket, wiring and ballast itself. The life expectancy of the entire unit is thus increased.

The LPS lamps are the most efficient lamps available today, delivering 137 to 183 lumens per watt depending upon lamp wattage. Continued development may enable low pressure sodium lamps to come closer to their theoretical maximum efficacy of 510 lumens per watt. LPS produces a monochromatic yellow light which tends to render all colors except yellow as shades of gray. Further, it is a linear source, making it difficult to control and distribute its light output efficiently. Although LPS is finding applications in this country, the lamps at present are all imported, which could lead to problems of availability or quality control.

PRESENT AND FUTURE LIGHTING TECHNOLOGY

Now we will turn for a moment to have a look at some of the newest breed of lighting developments to come on the scene and some which are about to be unveiled within the coming months. In the field of residential use, hotel rooms, lounges, lobbies and hallways, and private executive office suites, the following lighting sources will surely appear in conventional table lamps and decorative ceiling fixtures and see some limited use in various types of downlights. (See Figures 6–7, 6–8, and 6–9.)

Figure 6–7. Today's lighting designer producing the lighting solutions for tomorrow's residences. (Photograph courtesy of Manville Corporation *Future* magazine.)

Figure 6–8. Today's lighting designer producing the lighting solutions for tomorrow's retirement homes. (Photograph courtesy of Equity & Law Assurance Society Limited.)

A. The *Super Light Bulb* consists of a transparent coating of silver and titanium oxide [a Massachusetts Institute of Technology laboratory developed technology] applied to the inside surface of a glass incandescent bulb. This design feature will trap the heat inside the glass bulb in order to eliminate most of the 90% electrical energy usage that is lost as heat in a conventional incandescent light bulb. Duro-Test plans to test market these lamps this year. As a result, a 40-W superbulb will produce the same lumens available from a conventional 100-W incandescent bulb.

B. Some time this year GE's new *Electronic Halarc* lamp will be available to the public. The new lamp will be three times more efficient than any incandescent lamp, last up to four times longer and consume less electric energy while producing more light. It has a two-way 75- and 150-W setting and has a 5,000 hour life. It uses a ballast to start

Figure 6-9. Today's lighting techniques applied to lighting requirements of future underwater motels and recreational facilities. (Photograph courtesy of General Motors Futurama Exhibit.)

and regulate a metal halide high intensity discharge (HID) arc tube which is made of quartz and in addition has two incandescent filaments for providing instant light when first turned on.

C. GTE/Sylvania also has come up with a metal halide replacement for the incandescent lamp, called the *miniarc lamp*. This highly efficient 40-W unit produces more illumination than a standard 100-W incandescent lamp and has a rated lifetime of 7500 hours. The lamp with integral disposable ballast is only 7 inches long and 2.5 inches in diameter. Because of its compact size it can be used both in table lamps and in various other household fixtures. A lower power-consumption, high-frequency, electronic circuitry "ballast" controls the lamp. An electric discharge of an arc between two electrodes through the metal halide amalgam produces the light. Initial light output is low and an instant-on capability will be built into it, using one of several methods as yet to be decided upon.

D. Westinghouse, on the other hand, has developed a compact fluorescent replacement for the incandescent bulb. The light source is a $\frac{3}{4}$-inch diameter by 17-inch long fluorescent tube folded into a double U-shape that occupies an area 4 inches long by 2 inches in diameter. A plastic globe encloses the top and sides of the tube as well as encapsulating a large portion of the half-inch square by 3-inch long finger ballast. The enclosure gives the lamp a contemporary appearance. The overall size from the top to the base is 7.5 inches by 3.25 inches in diameter, which will allow it to fit inside the harp of most table lamps. The compact fluorescent lamp and ballast consumes only 27 watts of energy while producing 1000 lumens. The fluorescent lamps will last 7.5 to 9 times longer than an incandescent lamp with a lifespan of 7500 hours rating on 3 hours per start, or 15,000 hours of life on continuous use. The fluorescent lamp plugs into the lower ballast and base in order to allow lamp replacement since the ballast will last much longer and will outlive several replacement tubes.

E. Not to be outdone, GE has been evaluating a rather costly solenoidal-electrodeless fluorescent light source. This unit uses a high frequency ballast in conjunction with a ferrite ring that replaces the electrodes of a traditional fluorescent source. It, too, fits the standard household socket and produces about the same light as a 150-W incandescent lamp but at half the energy use.

F. In addition, GE is studying an infrared (IR) lamp which has a specially designed spherical bulb with an inside coating to reflect the heat back in and thereby increase its percentage of the energy converted to light. Its efficacy and design are similar to the approach being taken by Duro-Test Corporation.

G. G.E. self-ballasted, circular, flourescent lamps for insertion into table lamps or some large overhead fixtures have a lamp life of over 7000 hours and ballast lives listed at 50,000 hours, and are already commercially available in a 44-W unit. It has the light output of a 100-W incandescent lamp. Future variations will come in assorted wattage sizes and in various colors which will become available due to the use of different lamp phosphors.

H. GTE/Sylvania has just introduced a new line of incandescent and Tungsten-Halogen lamps utilizing a krypton gas-fill. This inert gas-fill allows the lamp filament to operate at higher temperatures and at higher efficacies. This new breed of lamps is called *Super Saver* lamps by Sylvania. The new incandescent lamps are from 10 to 15 percent more energy efficient than their standard counterparts and come in the following wattage sizes: 34, 52, 67, 90 and 135. They are intended as replacements for the 40-, 60-, 75-, 100- and 150-W standard long life incandescent lamps, respectively. The new lamps have a rated life of 2,500 hours and over their lifespan can save an owner anywhere from 90¢ using the new 34-W lamp, $1.20 with the new 52- or 67-W lamps, $1.50 with the new 90-W lamp and $2.25 with the 135-W lamp, based upon 6¢ per kWh. These newer krypton gas-filled incandescent lamps cost $1.44 per bulb for each of the various wattages. The 67-W Super Saver lamp is 12% more lumen efficient than the standard 75-W incandescent at an initial lumen rating of 970. This represents an efficacy of 14.5 lumens per watt (lm/W) versus 12.9 lm/W. The 90-W Super Saver lamp has an efficacy of 16.2 lm/W versus the conventional 100-W incandescent lamp at 14.6 lm/W that it is intended to replace.

1. The new 400-W Tungsten-Halogen lamp is 12.5% more lumen efficient than its previous 500-W counterpart at 22.5 lm/W versus 20 lm/W. The new 1200-W Tungsten-Halogen lamp is 25.38% more lumen efficient than the 1500-W unit it replaces, at 24.2 lm/W versus 19.3 lm/W. The krypton-filled 400-W Tungsten-Halogen lamp costs $32 versus $27 for the conventional 500-W lamp it replaces. At 6¢ per kWh and a rated life of 2000 hours, it saves $12 over its lifespan to more than defray the additional $5 first cost. The krypton-filled 1200-W Tungsten-Halogen lamp costs $45 versus $38 for the conventional 1500-W lamp it replaces. Over its 2000 hours life it saves $36 versus the initial $7 additional cost.

2. GTE recently announced new Sylvania Capsylite™ flood and spot lamps that will reduce energy by 40%. These 90-watt PAR lamps produce the same illumination as standard 150-watt PAR flood and spot units. The new Capsylite lamps consist of three basic parts: a halogen capsule, reflector and lens. The hard glass capsule, similar to halogen capsules introduced to the automotive market by GTE in 1979, produce a whiter, more efficient light than standard tungsten filaments. The capsule is precisely positioned within the reflector to maximize light output. Lamp performance is further optimized by the lens which concentrates the light in either a flood or spot pattern virtually eliminating wasteful fanning of light energy. PAR lamps are

popular for many commercial and residential applications such as displays, signs, facades, walkways and security lighting.

LATEST CURRENT TECHNOLOGICAL DEVELOPMENTS

We will now delve into the latest developments in current technologically updated lighting sources and their controls and end up with an in-depth analysis of a current trend (task-ambient, in-furniture lighting systems) that seems to be bidding us farewell after only a brief and transient appearance in the lighting arena.

LAMPS

Straight, 40-watt, rapid-start (RS) fluorescent lamps in the cool-white series have 3150 lumens (78.75 lumens per watt efficacy). (See Figures 6–5 and 6–6.) These lamps are normally rated for about 20,000 hours' life at three hours per start; 28,860 hours' life at 12 hours per start, and 37,700 hours' life on continuous burning, at a cost of $2.30 each (fluorescent lamp prices are identical for the three major manufacturers). (See Figure 6–10(A).) They are the least expensive and longest lived lamps in general use and are the most common interior lighting lamps. A 4-lamp 2 × 4′ fixture using 40-watt lamps consumes 184 watts, including ballast losses. Another variation of the straight, 40-watt, rapid-start fluorescent lamp is the U-shaped, 40-watt, rapid-start tube. This lamp and its attendant ballast and lighting fixture, along with operating and life-cycle costs are less expensive in the 2-lamp, 40-watt version than in the 4-lamp, 20-watt, trigger-start lighting fixture.

The following cost model compares first and life-cycle economic cost and energy use of U-shaped, 40-watt, RS lamps versus 20-watt trigger-start lamps.

Limited use should be made of 2′ × 2′ fixtures where required for an even lighting pattern, such as in corridor spaces and where a column would interfere with the uniform orderly installation of a standard 2′ × 4′ fixture. These fixtures should employ two U-shaped 34-, 35- or 40-watt fluorescent lamps rather than four 20-watt, trigger-start lamps. Lamp analysis follows:

U-shaped, curved, 40-watt fluorescent lamps of the cool white series have 2,950 lumens (73.75 lumens per watt efficacy). They have

MANUFACTURER	TRADE NAME	NOMINAL LAMP WATTS	INITIAL LAMP LUMEN	INITIAL LAMP EFFICACY (LUMENS/WATT)	LAMP COST	COST DIFFERENTIAL IN CENTS
GENERAL ELECTRIC	STANDARD COOL WHITE RAPID START	40	3150	78.75	$2.30	
	WATT-MISER	34	2850	83.82	$2.75	45¢ (vs. $2.30)
	LITE WHITE II	34	3050	89.71	$3.20	90¢ (vs. $2.30)
WESTINGHOUSE	STANDARD COOL WHITE RAPID START	40	3150	78.75	$2.30	
	COOL WHITE ECON-O-WHITE	34	2800	82.35	$2.75	45¢ (vs. $2.30)
	ECON-O-WHITE ECON-O-WATT	34	3050	89.71	$3.20	90¢ (vs. $2.30)
SYLVANIA	STANDARD COOL WHITE RAPID START	40	3150	78.75	$2.30	
	SUPER SAVER	34	2850	83.82	$2.75	45¢ (vs. $2.30)
	LITE WHITE SUPER SAVER II	34	3050	89.71	$3.20	90¢ (vs. $2.30)

Figure 6–10(A).

an 18,000-hour life. The ballast and fluorescent-tube power consumption are identical to a single- or two-tube, straight, 40-watt, RS fluorescent lamp and ballast installation. These lamps cost $8.00 each and use the same ballast as the straight tube.

The straight, 20-watt, trigger-start fluorescent lamps in the cool white series have only 1,260 lumens per lamp (63 lumens per watt efficacy). They have a 9,000 hour life, depending upon the characteristics of the starter. A single tube, lamp and ballast installation consumes 25 watts of power. These lamps cost $3.00 each.

A comparison of a 2′ × 2′ fluorescent lighting fixture having two 40-watt U-shaped fluorescent lamps versus the same fixture with four 20-watt, trigger-start lamps may be made.

A 2-lamp, U-shaped, 40-watt, rapid-start fluorescent lamp fixture has:

- 5,900 total lumens
- 92 watts of power consumed
- 18,000 hours of life
- $16.00 worth of fluorescent tubes.

A 4-lamp, 20-watt, trigger-start fluorescent lamp fixture has:

- 5,040 total lumens
- 100 watts of power consumed
- 9,000 hours of life
- $12.00 worth of fluorescent tubes

$$(\$12.00) \times \frac{(18{,}000 \text{ hours of life})}{(9{,}000 \text{ hours of life})} = \$24.00 \text{ worth of comparable lamps in the same time span}$$

In the 4-lamp, 20-watt fluorescent fixture without consideration of the cost of labor to install the 20-watt trigger-start lamps more often:

$$\frac{(5{,}900 \text{ lumens}/2\text{–}40\ W\,RS)}{(5{,}040 \text{ lumens}/4\text{–}20\ W\,RS)} \times \frac{(100 \text{ watts}/4\text{–}20\ W\,RS)}{(92 \text{ watts}/2\text{–}40\ W\,RS)} = 1.2724,$$

which means that there are 27.24% more lumens for every watt of energy consumed by the ballast and tube combination for the 2-lamp, 40-watt U-shaped fluorescent fixtures over the 4-lamp, 20 W, trigger-start fluorescent fixture. The U-shaped 4-foot long lamp now comes in a 34- and a 35-watt version which gives nearly the same light output (2950 lm/W for the 40-watt vs. 2650 lm/W for Sylvania's 34-watt, U-shaped lamp) but with considerable energy savings over the life of the lamp and its lighting fixture. GE's 35-watt, U-shaped lamp has 2600 lumens versus its 40-watt, U-shaped lamp at 2825 lumens.

The newer low-wattage, 34-watt, energy-saving, 4-foot long, rapid-start lamps have efficacies between 83.8 and 89.7 lumens per watt (varying according to manufacturer) (See Figure 6–5 and 6–6) and rated lives similar to the 40-watt lamp. The higher efficiency lamps reduce energy use while maintaining nearly the same illumination level and cost $0.90 more than a standard, 40-watt, rapid-start lamp. Over a 20,000 hour life, each low wattage lamp represents a savings of $7.20 based upon an energy cost of 6¢/kWh. At a normal 60 hours' use per week, energy-saving lamps justify their increased cost within 9.6 months.

Where reduction in illumination levels of 9.69% can be tolerated, another lower efficacy energy saving lamp can be used that costs only $0.45 more than a standard 40-watt lamp based on retail pricing. These lamps have been available since 1976 and produce 83.8 lm/W. The energy savings are identical to the more expensive lamp, but because of its lower first cost this lamp justifies its extra cost within 5 months at 60 hours' use per week. In practice these lamps when sold in bulk normally have a wholesale price half of retail and sell at the same price as conventional 40-watt lamps. Thus, any energy cost savings are accrued at no additional first cost. The low-cost, lower-efficacy, energy-saving lamp is particularly appropriate for retrofitting existing lighting installations. Since many lighting systems of the past 5 to 10 years were designed to deliver 70–150 fc, relamping with inexpensive low-wattage lamps offers a simple, effective way to cut operating costs while still meeting or exceeding today's lower illumination standards (currently, for example, NASA, the Navy and GSA require only 50 fc for office or school lighting.) (See Figures 6–10(A) and 6–10(B) for comparative lamp data.) Substantial annual energy savings can be realized by the use of these lamps as well as the spin-off air conditioning energy savings.

Of particular interest is the fact that the newer low wattage 34-watt lamps (either in the 2850 or 3050 lumen output range) may tend to

TYPE	LAMP	SOURCE	WATTS	SHAPE	MANUFACTURER	CATALOG NO.	INITIAL LUMENS	RATED LIFE	LAMP COST **
L1	T-12	F	40	TUBULAR	* GENERAL ELECTRIC SYLVANIA WESTINGHOUSE	F40LW/SB -------- F40T12/CW/1S/60	3450 ---- 3400	15,000 ------ 7,500	$2.40
L2	T-12	F	40	TUBULAR	* GENERAL ELECTRIC SYLVANIA WESTINGHOUSE	F40CW F40CW F40CW	3150 3150 3150	20,000 20,000 20,000	$1.39
L3	T-12	F	34	TUBULAR	* GENERAL ELECTRIC SYLVANIA WESTINGHOUSE	F40LW/RS/WM11 F40LW/RS/SS F40EW/RS/EW-11	3050 3050 3050	20,000 20,000 20,000	$2.11
L4	T-12	F	34	TUBULAR	* GENERAL ELECTRIC SYLVANIA WESTINGHOUSE	F40CW/RS/WM F40CW/RS/SS F40CW/RS/EW-11	2850 2850 2800	20,000 20,000 20,000	$1.39
L5	T-12	F	40	U-SHAPED	* GENERAL ELECTRIC SYLVANIA WESTINGHOUSE	F40CW/U/6 FB40/CW/6 FB40/CW/6	2900 2950 2950	12,000 18,000 12,000	$6.12
L6	T-12	F	35	U-SHAPED	* GENERAL ELECTRIC	F40CW/U6/WM	2600	12,000	$7.17

* PERFERRED LAMP SELECTION FROM THE STANDPOINT OF LAMP EFFICACY (HIGHEST LUMENS PER WATT), COLOR MATCH AND COLOR COMPATABILITY OF THE LAMP PHOSPHORS BETWEEN DIFFERENT LAMP TYPES.

** LAMP COST WHEN BOUGHT IN BULK QUANTITIES. PRICES REFLECT GENERAL ELECTRIC LAMP COST.

Figure 6–10(B). Lamp schedule.

decrease the life of a conventional 40-watt, RS ballast. The higher-efficacy, lower-wattage lamps tend to elevate the voltage impressed across the capacitor of the ballast and rupture it. In general six months out of a possible 13- to 15-year life may be lost. Some of the lamp and ballast manufacturers have speculated that if a 40-watt, RS ballast has been seasoned on 40-watt lamps for three or more years and then is relamped with the low wattage lamps, this may bring about a premature failure (loss of 6 months of ballast life). However, if the same ballast has been seasoned from the start with the higher efficacy, lower wattage lamps, then no premature failure should occur. Universal ballast manufacturers do not feel that either of the above two conditions will have any effect on their ballast or shorten its life expectancy.

The majority of the ballast and lamp manufacturers do feel, however, that if the existing ballast has an internally shorted capacitor in its ballast, then the insertion of the lower wattage lamps will hasten its demise almost immediately.

Another variation on the energy savings lamps, but a different approach, is Sylvania's *Thriftmate 33*. The Thriftmate 33 is a fluorescent lamp that incorporates an integral dimming unit which must be operated on a 2-lamp ballast in conjunction with a standard 40-watt or energy-saving RS lamp. This type of reduced energy lamp is not available in the U-shaped, 40-watt fluorescent series. The Thriftmate 33 used in conjunction with a standard, 40-watt, RS lamp on a 2-lamp, 40-watt ballast cuts the illumination and energy use by 33%.

Still another variation of this is the GTE/Sylvania *Thriftmate 50* which, in conjunction with a standard, 40-watt, RS lamp on a 2-lamp, 40-watt ballast, cuts illumination and energy use by 50%. In most cases if we are dealing with a 4-lamp lighting fixture it would be easier to delamp the outboard tubes and open-circuit the primary leads of the ballast. Merely delamping a ballast will allow it to continue to consume its 12 watts of energy needlessly and hasten its unnecessary demise. If we delamp all the lighting fixtures of 50% of their lamps with the ballast open-circuited and on the line, the power factor will deteriorate appreciably, increasing the current per usable working kW. This will, in turn, reduce system capacity, increase current flow and voltage drop, as well as increase I^2R line losses (creating additional heat losses) which will decrease the life of the electrical system and marginally increase the air conditioning requirements.

The Thriftmate 33 and 50 lamps create an uneven brightness pattern within the luminaire since the end that houses the dimmer produces no light. (See Figure 6–11) It should also be kept in mind that Sylvania has in the past recommended against using them in conjunction with 34-watt fluorescent lamps. Sylvania instructions state that the Thriftmate 33 and 50 are designed to operate only on two-lamp, 40-watt, series-sequence, rapid-start ballasts, with a standard, 40-watt, T-12 (F40) lamp in the other socket. Sylvania further warns that use with equipment other than as specified may result in overheating of the equipment.

Through continual on-board testing and research, GTE/Sylvania has found that the Thriftmate lamps in conjunction with their Super Saver lamps should not cause premature failure of the lamps or ballast. However, they have found that upon first starting up, the lamps will striate and flicker for several minutes as the arc plasma swirls within the lamp. This condition can be expected to last for several thousand hours after the initial installation of new lamps. If this condition is not found uncomfortable or undesirable by the occupants of the space, then it should not prove to be too great a liability.

California Institute of Technology invented a device that became known as the *phantom tube*. Developmental Sciences, Inc. developed and perfected the technology of a substitute lamp that emits no lumens of visible light itself but when placed in a 2-lamp, 40-watt, RS, fluorescent ballast circuit not only does not produce light; it does not consume the 40 watts of energy. This energy-saving approach creates an uneven brightness pattern within the luminaire since alternate lamps within it emit no light. In addition, it reduces the energy consumed by the ballast and its companion 40-watt lamp. Since 2-lamp, 40-watt, RS, fluorescent lamp ballasts are wired in series, the removal of only one lamp will extinguish the other lamp. In addition, an original power factor of from 0.96 (for low heat ballasts) at 277 volts to 0.98 PF for a conventional 277-volt ballast will, upon open-circuiting, be reduced down to 0.20 PF. Ballasts under such circumstances behave similarly to that of a non-loaded transformer having an inductive reactance and very low PF. The voltamp product of such

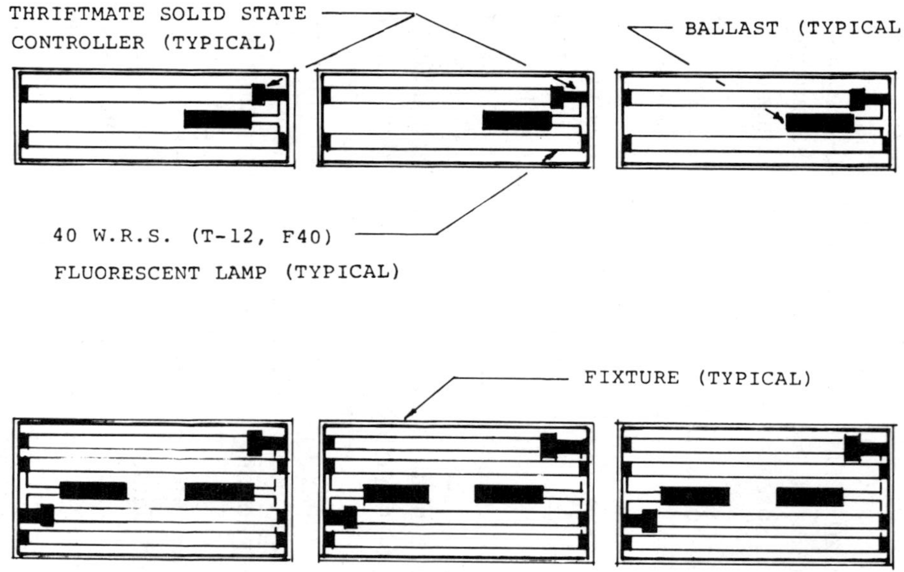

Figure 6–11. Several acceptable solutions to minimizing the possible distractions caused by the Thriftmate lamp end taper creating an uneven brightness pattern within the luminaire. (A) All two (2) tube, 40-watt fluorescent lighting fixtures located in the same room, corridor or area shall have the Thriftmate energy reducer lamp installed as follows: the Thriftmate solid state controller shall always be oriented on the same end and on the same side of the lighting fixture shown in the upper detail. (B) All four (4) tube, 40-watt fluorescent lighting fixtures located in the same room, corridor or area shall have the Thriftmate lamp installed as follows: the Thriftmate solid state controller shall retain the identical location and orientation from fixture to fixture as shown in the lower detail.

a ballast is approximately 45% of the normal consumption in a fully loaded 2-lamp ballast with lamps.

By inserting the phantom tube in lieu of the lamp, that lamp's 40 watts of energy are no longer consumed and the companion lamp and ballast also consume less power. The total energy consumption for the 2-lamp, 40-W lamps and ballast in practice is reduced down to a range of from 26.1% to 32.7%. Since the phantom tube device has additional capacitance, it allows the other remaining lamp to operate more efficiently at the reduced energy consumption level. With a reduction in energy use of from 63.9% to 67.3%, this might seem indicative of a reduction in light output of from 63.9% to 67.3% for a light output of only 26.1% to 32.7%. However, the remaining 40-watt lamp is operating more efficiently and at about 54% of lumen output. The 40-watt, RS version of the phantom tube retails at between $8 and $9 depending on location of store and distributor. If the lighting system is used 12 hours a day, 5 days a week and 52 weeks a year, its annual operating time will be 3120 hours. At an energy savings of 68 watts per phantom tube the yearly savings in energy will be (68 watts ÷ 1000 watts/kW) × 3120 hrs equals 212.16 kWh. At 6¢/kWh the accrued annual energy cost savings will be $12.73 for a payback period of from $7\frac{1}{2}$ to $8\frac{1}{2}$ months, based on the initial cost of the lamp only (no labor charge included.)

Each phantom tube paired in series with its respective size lamp will save the following amounts of energy as shown in Figure 6–12.

BALLASTS

Depending upon the manufacturer, low wattage ballasts save 3 to 5 watts (which can be a substantial savings for a large installation) and are compatible with 34-watt fluorescent lamps but will not necessarily properly operate 40-watt fluorescent lamps. Several ballast manufacturers such as Universal, manufacture a low-wattage ballast compatable to both the 40- and 34-watt energy-saving lamp that will properly start and operate these lamps. Low-loss ballasts such as GE Maxi Miser I, Advance Mark III, and Universal SLH are CBM approved for use with standard F40 lamps. However, when these same

LAMP MODEL REPLACED	KILOWATTS SAVED/TUBE
F20 LOW POWER FACTOR	.018
F20 HIGH POWER FACTOR	.030
F30	.050
F48 (INSTANT START)	.055
F40	.068
F96 (INSTANT START)	.104
F96HO	.176
F96VHO	.328

Figure 6–12. Energy savings with the phantom tube.

ballasts are used with energy saving lamps, the ballast factor is reduced to about 88%; meaning lumen output is lower than design levels. The GE Maxi Miser II ballast is designed to provide a 95% ballast factor where used with energy-saving lamps—typical of a standard full wattage system. The Maxi Miser II ballast however does not meet CBM standards and runs at a higher system wattage than other low loss ballasts. However, low-wattage ballasts cost from \$3.50 to \$5.50 apiece more than conventional ballasts (or \$7.00 to \$11.00 more for a 4-lamp, 2-ballast, 2′ × 4′ fixture), and the additional investment will not be amoritzed in less than 3.74 to 5.88 years, based on 60 hours' use per week and 52 weeks per year and 6¢ per kWh (the payback period does not take into account the interest that must be paid on the additional principal used to buy the more expensive-ballasts.) One manufacturer claims for its low wattage ballasts a life expectancy of from 2 to $2\frac{1}{2}$ times that for conventional core and coil fluorescent ballasts. Low wattage ballasts are primarily designed for use with low wattage, 4-foot lamps, but in their present state of development are considered compatible with the conventional 40-watt RS lamp as well. Lumen output for low-wattage lamps is generally based upon use with conventional 40-watt RS ballasts. The original, low wattage ballasts produce lower lumen outputs than advertised for lower wattage lamps; however, the G.E. Maxi Mizer II low wattage ballasts and lamps are designed to produce rated lumens. Standard 40-watt RS and very low heat (VLH) 40-watt RS ballasts will, without exception, allow the lower wattage, higher efficacy lamps to produce listed lumen ouput. In any case, conventional ballasts and VLH ballasts allow either 40-watt or 34-watt RS lamps to be used and at their rated output. A more practical option to using standard ballasts is to specify all ballasts be Very Low Heat Rise type, as they are compatible with 34-and 40-watt fluorescent lamps. The VLH ballast represents an additional initial \$3.50 cost over a more conventional ballast (or \$7.00 per 4-lamp, 2-ballast fixture). However, ballast manufacturers' figures indicate that even slight reductions in the continuous ambient operating temperature of a ballast will result in extension of ballast life by a substantial factor, thus increasing the maintainence-free life of the lighting installation and substantially reducing the overall cost of ownership of the system by reducing energy consumption. The ballast coils have conductors that are larger in cross-sectional area; thus their internal I^2R energy losses are lower, hence their cooler running performance and increased life span. Figure 6–13 compares the costs and characteristics of standard, low wattage and VLH ballasts.

Regardless of their other characteristics, fluorescent lamp ballasts should be the 2-lamp type, and single-lamp and single-lamp ballast fixtures should be minimized as much as possible. As Figure 6–12 indicates, a single-lamp plus ballast consumes 52 to 54 watts (using 40-watt RS lamps and depending upon manufacturer); and two lamps plus ballast, consume 92 to 94 watts (using 40-watt RS lamps). Thus, two single-lamp fixtures can consume as much as 108 watts or 17.4% more energy than a 2-lamp, single-ballast fixture at 92 watts. Where continuous rows of single-lamp fixtures must be used (such as in cove lighting), maximum energy can be saved by specifying the continuous fixtures as tandem units so that each pair of in-line lamps is operated by a 2-lamp ballast.

A solid-state electronic ballast has many potential advantages such as energy-saving efficiency, ability to be manually or automatically dimmed, small size, less weight, no visible flicker and no audible noise

BALLAST MFR.	POWER CONSUMED BY STANDARD CLASS P BALLASTS AND LAMPS (IN WATTS)				ADDITIONAL PRICE (OVER STANDARD BALLAST) FOR LOW-WATTAGE 2 LAMP BALLAST	POWER CONSUMER BY LOW-WATTAGE BALLAST ONLY (IN WATTS)	ENERGY SAVINGS WITH LOW-WATTAGE BALLASTS (IN WATTS)
	BALLAST ONLY (1 LAMP)	BALLAST & LAMP COMBINED 1/40	BALLAST & LAMP COMBINED 2/40	BALLAST ONLY (2 LAMP)			
GENERAL ELECTRIC	12	52	94	14	$3.50 MORE	8-9 WATTS	5 TO 6
UNIVERSAL	12 14	52 54	92 94	12 14	$3.50 MORE	7-9 WATTS	3 TO 5 5 TO 7
JEFFERSON	12	52	94	14	$5.50 MORE	NOT AVAIL.	---

Figure 6–13. Cost and energy savings comparison between conventional standard and energy saving ballasts.

to name but a few. There may also be many as yet unknown disadvantages and those that are partially known or understood are incomplete at best. Extension of lamp life and ballast life has yet to be proven and electromagnetic (EMI) and radio frequency interference (RFI) may be greater and therefore more deleterious in some instances than the more conventional core and coil ballast. Lamp life is based upon a magnetic ballast delivering all the electrical characteristics to the lamp in accordance with ANSI specifications. There are no current ANSI specifications for electronic ballasts. Thus there are no parameters and no way to judge if the electronic ballasts meet or exceed standard magnetic ballast performance except for speculation and educated guesses. Therefore, extended lamp-life inference is only a guess. Ballast life is based primarily on componenet reliability of accumulated quality control based upon the quantity and quality of components. Depending upon manufacture of typical low-energy magnetic ballasts there are only 3 to 4 components in them. Of the 15 or more manufacturers of "solid-state" electronic ballasts there are from 21 to 87 total components with from 3 to 9 magnetic ones.

Electronic ballasts have higher efficiencies by operating the fluorescent lamps—from 10% to 14% greater efficiency when energized in the 19 to 28 kiloHertz AC frequency range. The ballast itself is more efficient than conventional core and coil magnetic ballasts in that it consumes 4 watts versus 12 watts. Whereas 9 watts are consumed in a premium energy saving magnetic ballast. The combination of reduced ballast energy consumption plus a lower wattage utilization by the lamps while producing approximately the same lumen output decreases the heat build-up in a lensed recessed grid troffer thus reducing thermal lamp lumen depreciation which further increases the combination of the lamp and lighting fixture efficiency.

There is little or no visible lamp flicker since 22 kiloHertz electronic ballast cycles the lamp on and off 44,000 times per second as opposed to 120 times per sec for a conventional household alternating current of 60 Hertz.

A 2-lamp, 40-watt, rapid-start, fluorescent-lamp, electronically-ballasted fixture uses about 13 watts less energy than an energy-saving, magnetic, 2-lamp ballast. An additional $25 per OEM-installed electronic ballast in new lighting fixtures can be expected above the base cost for an energy-saving magnetic ballast.

For a 12 hour a day, 5 day a week operation, 3120 hours of fixture operation will be required annually, at a cost of 6¢ per kWh. The energy cost savings will be $2.43 per lighting fixture requiring a 10.27 year payback for the electronic ballast.

Sylvania has experimented with and recently perfected an electronic ballast but is no longer in the electronic ballast business. This 60-Hertz input ballast has an inverter that produces a much higher frequency output that gives a 12% improvement in lamp and ballast efficiency. The higher frequency ballast output produces more visible energy from the mercury atoms and excites the fluorescent lamp phosphors to higher levels of visible energy output so as to produce a 12% to 15% greater lamp efficiency. Thus, for a given wattage lamp, the lamp's intrinsic lamp efficacy is improved, i.e., for the same energy usage there is more visible light output.

Recently Sylvania developed a new fluorescent lamp system using a special ballast to operate its new octron lamp. That ballast is now available from the Universal Manufacturing Company and Advance Transformer Company and is to be used in conjunction with two 4-foot long 32-watt lamps. Its total energy consumption is 24.1% lower as compared to 2-lamp, 40 W, RS lamps and ballast combined, while producing the same light output in an enclosed fixture.

The *Octron,* a new 4-foot long 32-watt lamp by Sylvania, is referred to by them as the "Fluorescent (lamp) of the Future." With four Octron lamps and two special ballasts to operate these lamps, 42 watts of energy can be conserved compared to a conventional 4-lamp, 40-watt, RS lamp fixture using two 2-lamp, 40-watt lamp conventional ballasts. This represents a 24.1% savings in energy for light (or 36% total savings including air conditioning) while still producing the same light output. (See Figure 6–14.) In addition, an improved color rendering index is obtained. These lamps come in two popular color temperatures (3100°K for warm tones and 4100°K for cool tones), while each produces a 75 CRI value. At 90 lumens per watt versus a conventional 40-W RS lamp at 78.75 lm/W, the energy saving costs on the improved lamps and ballast can repay the additional cost of this system within one year.

General Electric has just come out with a new ballast that is a combination of some of the old electromagnetic components wed with new solid-state circuitry. This high efficiency 6-watt ballast specially

SYSTEM	RELATIVE FIXTURE OUTPUT	TOTAL WATTS	RELATIVE SYSTEM EFFICIENCY	ANNUAL ENERGY COST*
ORDINARY LAMPS STANDARD BALLASTS	100%	174	100%	$31.32
F40 LITE WHITE ENERGY-SAVING LAMPS ENERGY-SAVING BALLASTS	95%	138	120%	$26.15
OCTRON LAMPS 32W T8 RS BALLASTS	100%	132	132%	$23.76

*BASED ON 3000 HOURS PER @ $0.06/KWH ADJUSTED TO 100% RELATIVE LIGHT OUTPUT

COMPARATIVE SYSTEM PERFORMANCE
(TYPICAL 4-LAMP LENSED TROFFER)

LAMP	PHOSPHOR	CRI
ENERGY-SAVING 4 FOOT FLUORESCENT	LITE WHITE DELUX LITE WHITE	67 48
ORDINARY F40 4 FOOT FLUORESCENT	COOL WHITE WARM WHITE	62 52
OCTRON F032 4 FOOT FLUORESCENT	OCTRON 3100K OCTRON 4100K	75 75

COMPARATIVE COLOR RENDERING INDEXES (CRI)

MANUFACTURERS NAME	ORDERING ABBREVIATION	4 FOOT LONG OCTRON FLUORESCENT WATTS LAMP	COLOR TEMP°	APPROXIMATE LUMENS	AVERAGE LIFE HRS
GTE SYLVANIA	F032/31K	32	3100K (WARM)	2900	20,000
GTE SYLVANIA	F032/41K	32	4100K (COOL)	2900	20,000

LAMP CHARACTERISTICS

Figure 6–14. Sylvania Octron fluorescents.

controls two 28-watt, 4-foot RS T–12 bipin fluorescent lamps, and is known as the *optimizer system*. The new General Electric indoor fluorescent lighting system is on the cutting edge of technology in energy-efficient fluorescent lighting systems. With a pair of optimizer ballasts and a complement of 4 lamps, it will use 60 watts less energy than a comparable 4–lamp, 40-watt system with standard ballasts. The new technology ballast, working together with the latest in technology 28-watt lamps, produces the lumen output comparable to that of today's typical energy-saving fluorescent lighting systems. The optimizer's system of matched components developed by one in-house organization helps boost the efficiency of the system 39%. The ballast's better performance is attributable to a unique solid-state switch that cuts off all power to the lamp's bimodal cathodes after lamp start. This feature reduces energy input without any loss of light output.

Thomas Industries, Inc., is a lighting fixture manufacturer that has turned to research, development and now production of another new concept in electronic ballast design. Their approach is a total package consisting of a lighting fixture supplied with cooler-operating, dependable, high frequency, higher efficiency electronic ballasts. This ballast is called the *watt watcher*. It comes only in 120 volts and is 50% lighter than the conventional electromagnetic ballast while providing more efficient conversion of electrical energy into lighting energy. The ballast, while providing nominal full light output for 34- and 40-watt RS lamps, reduces energy consumption 30% compared to a more conventional 2-lamp or 4-lamp, 40-watt RS lighting fixture. This imparts cooler operation for both ballast and lamps, thus extending their useful life and further increasing the lamp's optimum efficacy. The higher frequency generator (inverter) increases the lumens per watt output of the lamp and provides superior regulation of power consumption and light output under input line voltage fluctuations. The following is an energy savings performance comparison between a conventional standard 2-lamp, 40-W RS lamp and ballast at 88 input watts, a watt watcher with 2-lamp, 40-W RS lamps at 70 watts input and a watt watcher with 2-lamp, 34-W RS lamps at 60 watts input. The Thomas Industries test data show that with conventional lamps and a watt watcher ballast 18 watts or 20.45% of the energy is saved. A watt watcher ballast and 34-watt lamp uses 28 fewer watts or 31.8% less energy than the conventional lamp and ballast fixture. A careful scrutiny quickly shows that the watt watcher ballast plus 2-lamp, 40-watt lamps have only 70 input watts for 80 watts of connected lamp load plus ballast. Since the ballast is providing a higher frequency the lamps are running at a 30% increase in efficacy; therefore, in order to obtain the same light output as the conventional lamp ballast pair, a reduction in input energy can be tolerated. In addition, these ballasts have a PF in excess of 95%, and although they sell for $30 wholesale, when sold OEM-installed in the Thomas Industries' lighting fixtures, they cost around $14 apiece. A typical yearly operating period of 4000 hours with an energy cost of 6¢/kWh will save $6.72 annually for a payback period of approximately two years.

Alkco just recently pioneered a dimmable fast starting, high power factor, Class P, 120-volt, solid-state ballast for use in both under-shelf local task lighting and for headwall mounting for bedlighting in dormatories, nursing homes and military barracks. The dimmer has an integral potentiometer dimming switch and the unit controls two 13-watt T5 fluorescent lamps. The on/off potentiometer dimmer switch unit controls light output from 25% to 100% output.

HIGH PRESSURE SODIUM AND SUPER METAL HALIDE

High pressure sodium lamps offer the highest efficacy of any source used for general lighting of up to 140 lumens/watt. (See Figures 6–5 and 6–6.) (Low pressure sodium has considerably higher efficacies (183), but due to its objectionable monochromatic yellow color it's seldom used in interior applications.) HPS has, in the past, generally been relegated to street lighting applications due to the golden color of its light which rendered colors and complexions poorly. In some commercial applications, though, conventional HPS has been successfully used for interior lighting. By illuminating to 100 fc or more and carefully coordinating furnishings to use colors that are enhanced by HPS lighting, the psychological impression of white light can be closely approximated.

Also, over the last three years, HPS lamps with increased vapor pressures have been developed that produce a nearly white light comparable with warm white fluorescent and phosphor coated metal halide lamps but at a lower color temperature at an efficacy of 91, 95 and 100 lumens per watt, respectively, for the 150, 200 and 250 watt high CRI HPS lamps, which is slightly better than that of fluorescent lamps. These improved, or high CRI, HPS lamps will be available from all major lamp manufacturers so that, for the first time, HPS is a viable selection for interior lighting; in fact, HPS and SMH promise to be the preeminent lighting sources of the next decade. Interior lighting with 150-, 200- or 250-watt HPS fixtures permits an ideal mounting height-to-fixture spacing ratio of 1:1.4. A variety of lenses can be used with HPS fixtures if a glass optical interface is used to disperse the heat that could otherwise damage standard acrylic lenses.

An alternate approach is to light general areas with 400-watt SMH fixtures equipped with HID acrylic refractors that offer a mounting height-to-fixture spacing ratio of 1:1.9, far exceeding spacing ratios normal with fluorescent fixtures. SMH lamps provide excellent color rendition at an efficacy of 100 lumens/watt, and the latest HID acrylic refractor produces a flattened or *softened* batwing distribution of light, permitting a wide, even pattern of illumination without the problem of fixture brightness often noted with conventional linear or spherical batwing lenses used on fluorescent fixtures. The refractor also obviates the problem of hotspotting that would otherwise be caused by using high wattage lamps. Fewer fixtures are required using 400-watt SMH in place of 150-, 200- or 250-W HPS, but light distribution and illumination in most low mounting height applications would be marginally less even, especially if bank-style, open-office, architectural-space planning partitions are used. These will tend to cause unacceptable shadowing with widely-spaced fixtures.

AUTOMATIC ENERGY CONTROL (AEC)

Fluorescent lighting sources and HID sources, such as MV, MH, SMH and HPS lamps, are rated at initial light output. If the maintenance factor were not considered when establishing a maintained illumination level, a lighting installation designed for a particular footcandle level would start at that value only initially; the illumination would then steadily decrease from the day of installation. This is due to lamp lumen depreciation (LLD) and lamp and luminaire dirt depreciation (LDD) which together represent the maintenance factor (MF = LLD × LDD). The maintained footcandle level equals the initial illumination times the MF. Thus, when using lamps and lighting fixtures rated at 0.7 maintenance factor, a typical lighting system intended to deliver 70 fc is designed to deliver an initial 100 fc. Over the life of the lamps, such a system will provide an average 70 fc but will provide an actual 70 fc for only part of its life. However, the energy consumed over the period of usage is 100% of the connected load whether it produces 100 fc, 70 fc or 50 fc, (See Figure 6–15.)

With an Automatic Energy Control (AEC) system, available from several manufacturers, an initial illumination level of 100 fc is designed in order to maintain 70 fc during the life cycle. Only 70% of the energy is utilized initially to produce 70% of the initial illumination, or 70 fc. As the system ages, more energy is required until, many thousands of hours later, the full 100% energy usage is required to maintain 70 fc. However, the difference in energy usage over those thousands of hours represents a substantial cost saving while maintaining the desired 70 fc throughout the installation's life.

With AEC, fluorescent and HID lamps can be utilized to the end of their life cycle: 37,700 hours for fluorescent, 20,000 hours for horizontal burning SMH, 24,000 hours for HPS). (See Figures 6–16 and 6–17). However, a conventional fluorescent lighting system would be relamped at 75% of maintained lumens; SMH, at 70% of maintained lumens; and HPS, at 90% (for relamping at approximately 16,000 to 18,000 hours for fluorescent, 14,000 hours for SMH and 21,000 to 22,000 hours for HPS). The AEC controlled fluorescent or HID dimming system will thus reduce the maintenance cost attributable to more frequent labor and materials charges required by a conventional non-dimmed fluorescent or HID lighting system.

AEC works by controlling an automatic dimming system with a photoelectric control which senses available light in the room. Set to sense 70 fc, the photoelectric control, in addition to reducing the output of a new (initial 100 fc available) system to 70 fc, will also take into account available daylight. Hence, if the daylight contribution to a room is 21 fc, the light output of the fluorescent or HID fixtures, already operating at only 70% of full capacity, will then be reduced an additional 30%, providing a full maintained 70 fc at 49% energy consumption for the HID system (See Figure 6–17) and 39.2% energy consumption for the 20,000 Hertz fluorescent system (See Figure 6–16).

However, at this point in their development, high CRI HPS lamps generally should not be dimmed below 80% of full light output because of color shifting. Below 70%, they begin to lose their high

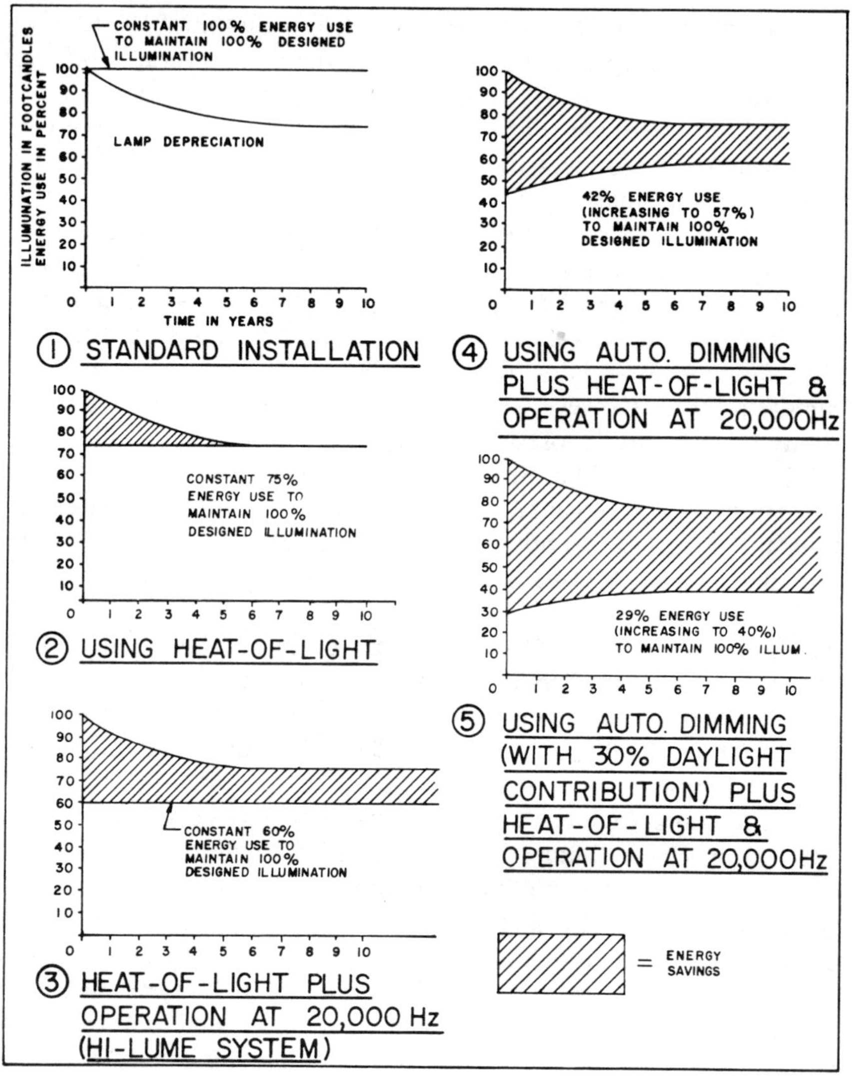

Figure 6–15. Energy savings-combining several design techniques.

color rendering property and at 60% exhibit the CRI (light color) of a conventional HPS lamp. When dimmed to as low as 45%, they produce a monochromatic light output very similar to low pressure sodium.

If fluorescent office lighting is used, optimum energy conservation

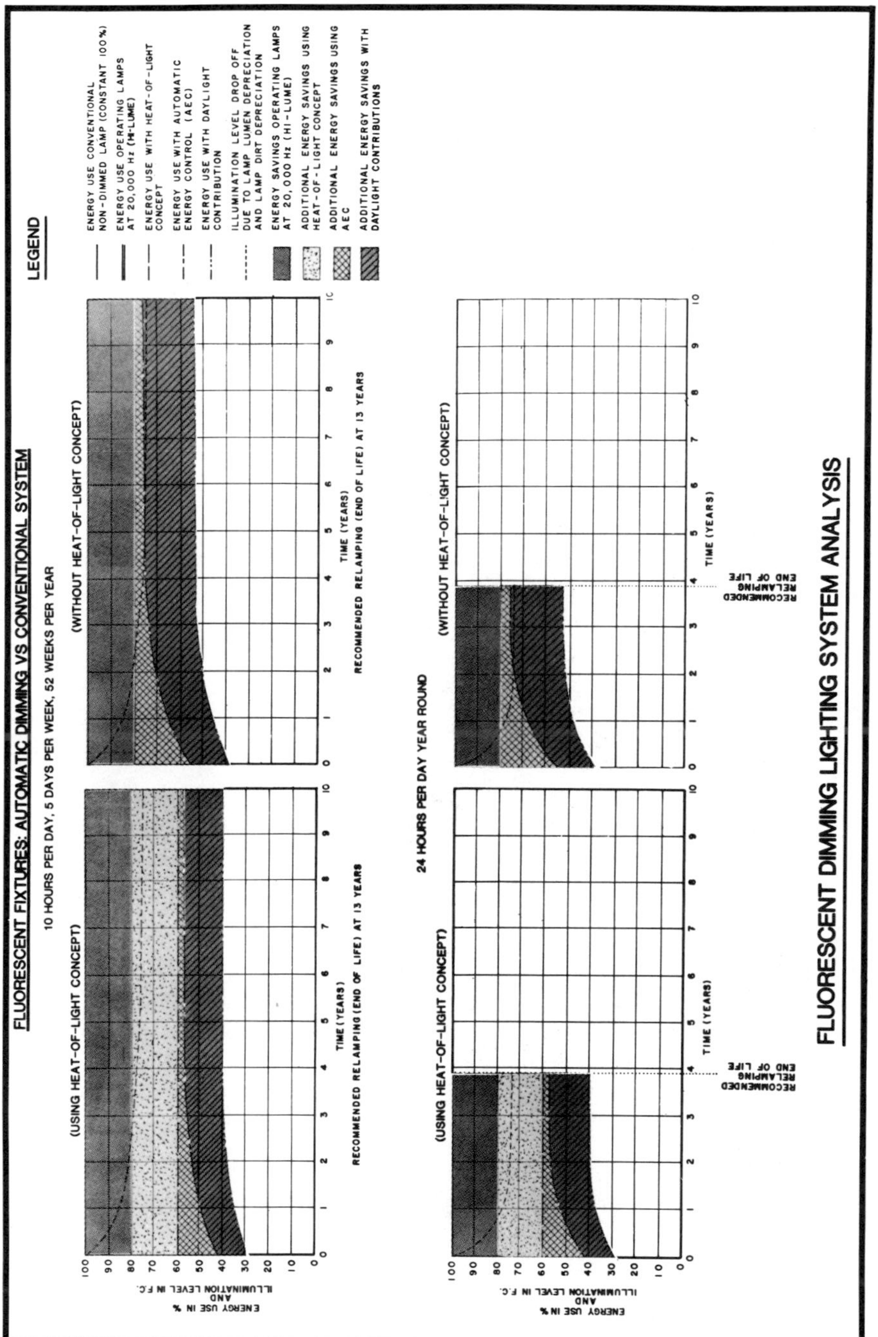

Figure 6–16. Fluorescent dimming lighting system analysis.

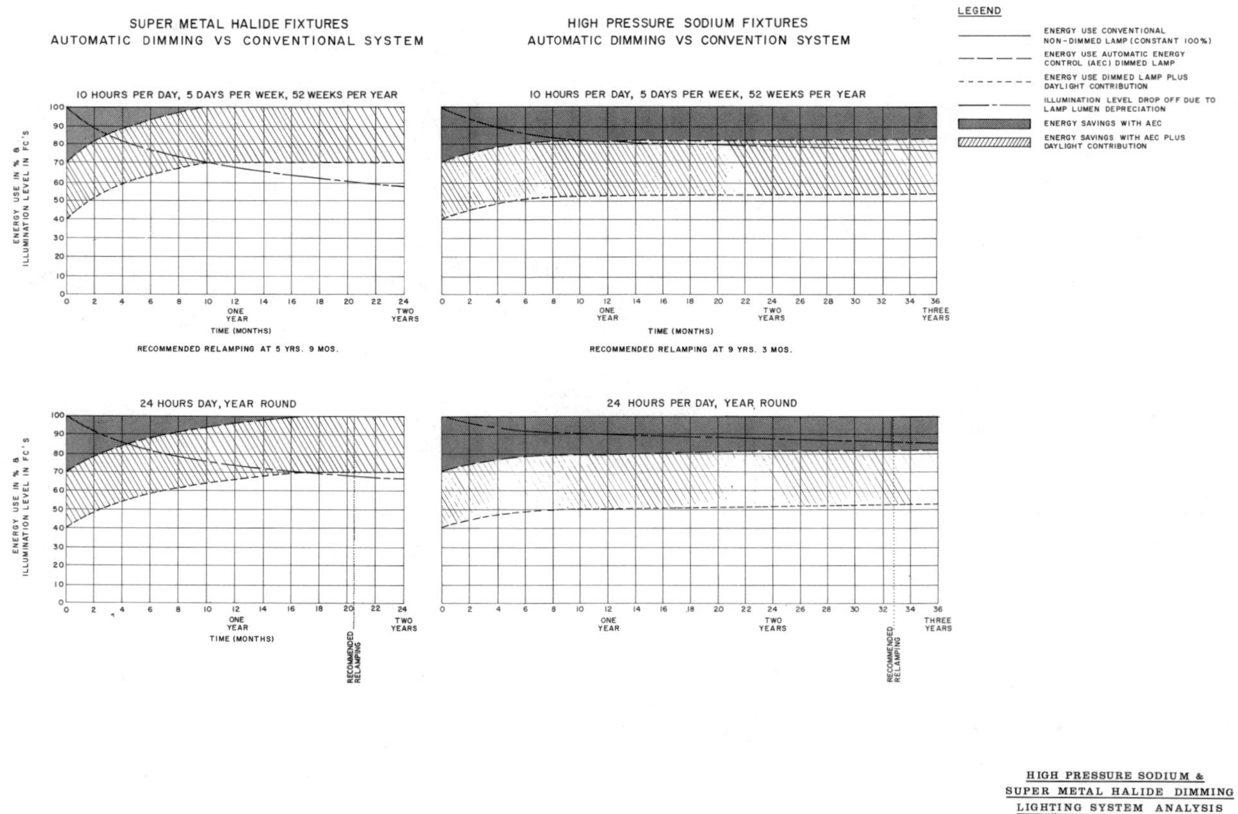

Figure 6–17. High pressure sodium and super metal halide dimming system analysis.

and lamp life can be realized through the use of a 20,000 Hertz dimming system. Operating fixtures at 20,000 Hz rather than the conventional 60 Hz extends lamp life (thus reducing maintenance costs) by 20%. A 20,000 Hz system also uses only 80% of the electrical energy of a conventional non-dimmed fluorescent system in order to produce 100% of its lumen output, whether or not the fixtures are dimmed. (See Figures 6–15 and 6–16.)

Although the above systems are descriptive of Lutron, Hub and Wide Light, there are several other different approaches that can be taken. Ashlar Industries, Inc., Enertron, Inc., and Lutron offer fluorescent dimmers that can be retrofitted to existing circuits in existing buildings without the need to replace the standard ballast with a dimming ballast. These retrofit systems are connected ahead of the circuit, that is, the system can be installed in the electrical closet adjacent to the lighting panelboard, or can be installed in the dropped ceiling cavity space ahead of all the fluorescent lighting fixtures on the circuit. With the aid of a photocell, the circuit dimming device lowers the light output from all the fluorescent lighting fixtures on that respective circuit. The retrofit and new system's equipment cost about an additional $5.00 per lamp installed.

There is a drawback to this system. If you turn the lights off in your room via local switching, you must bring all the fluorescent fixtures on that circuit up to full intensity (light output) prior to being able to turn your lights back on. The reduced voltage (potential gradient) and reduced arc current are insufficient in the dimming mode to allow the fluorescent lamp to restrike its arc.

Lutron has just recently come out with a new system that overcomes this initial starting difficulty. After the lights have been turned off in your room, upon their being turned on again (restruck) the overall system temporarily returns to full light output so that these lights can restrike and after 5 seconds return to their original level of dimmed output.

The Lumecon Fluorescent light dimmer by Ashlar Industries, Inc., not only can dim an entire 20-amp branch circuit, it can work with any standard 40-W ballast fluorescent lighting fixture. It works equally well with fluorescents as with incandescent lamp sources and lighting fixtures. The manufacturer claims substantial energy savings can be achieved with this dimming system and that a 30% reduction in lighting output corresponds with an energy savings of from 30% to 40%.

With the savings in energy costs this dimming system can be expected to pay for itself in just a little over a year. Its expected useful life is in excess of 20 years.

Another system on the market that is similar to the 20,000 Hertz Lutron solid-state electronic ballast and dimming system that can be utilized in conjunction with an AEC system is *Luminoptics*, which is a division of the Universal Manufacturing Corp. This system is designed to provide a high degree of flexibility and control while minimizing energy consumption. The Luminoptics solid-state, high frequency, inverter ballast system produces 27,000 Hertz output that causes fluorescent lamps to produce more output lumens with less energy input which in turn increases the lamp's efficacy. This, combined with a more efficient ballast that uses very little energy and can produce optimum light output from the lamps at reduced levels of energy use, substantially reduces total energy use below comparable levels of several other manufacturers. Its actual energy savings with this ballast is 45% more, as compared to the best core and coil ballasts, and is reported to be 25% more energy efficient than other solid-state ballasts currently on the market. Money saved by this more energy efficient system can repay the additional cost in less than a year. Since the ballast uses less energy it runs cooler and its projected life span is 55 years, based upon 4000 hours of use per year, which is two to three times longer than a conventional ballast. These ballasts are also interchangeable with conventional ballasts and do not preclude being mixed and matched with other manufacturers' ballasts. In addition, they have a PF of 90% or better and have less third harmonic distortion than a conventional core and coil ballast while maintaining a lower emission of electromagnetic interference.

New lighting standards require that lighting levels be calculated based upon the required speed and accuracy of the task, the age of the worker and visibility demands of the task. Regardless of whether the project is a retrofit or a rehabilitated structure, or even a new facility in which future tenant task requirements are unknown, the lighting levels can be custom tailored. A simple manual adjustment to each ballast adjusts the lamp lumen output necessary to match design requirements. This adjustment can be performed after a task function has been defined. It can also be used to compensate lighting levels in order to maintain worker productivity or to accommodate for worker age or difficulty in the visual task requiring higher levels of illumination. By the same token, the ability to trim individual ballasts and lighting fixtures for specific job requirements affords the flexibility of possible individual task lighting control.

When the Luminoptics solid-state, high-frequency ballast system is used in conjunction with an AEC system coordinated with an optical feedback from a zoned photocell, much greater savings can be accrued. When the solid-state, fluorescent ballast is used in conjunction with dimming equipment such as an AEC system, another very important feature was incorporated. Filament power is increased when the lamps are dimmed in order to insure longevity of the lamp. Under non-dimmed conditions, normal lamp filament power is maintained. This ballast is not only available for fluorescents but also for HPS lamps which can also be dimmed. Taking daylight contributions into account, along with adjusting local lighting levels for task requirements and dimming to preselected lighting minimums, fluorescent lighting systems can realize up to 70% in energy savings. HPS lamp lighting systems can realize upwards of 50% in energy savings, as well as related operating cost savings.

Individually automated, fluorescent lighting-fixture, dimming devices are now available through Ecalo by Controlled Environment Systems, Inc. *Ecalo* is an acronym for Energy Conserving Automatic Light Output and is a system that allows standard fluorescent lamps and ballasts to be used while conserving up to 60% or more electrical energy as is traditionally used in the same fluorescent lamp fixture. This system, like the previous AEC systems discussed, takes into account dust collection on the lamps, lenses and fixtures. It also accounts for gradually reduced lumen output due to cathode-emitted material coating the ends of the lamps (covering some of the lamp's phosphor molecules). In addition, LLD is also caused by some of the more energetic ultraviolet radiation in the electric arc destroying some of the phosphor molecules. Ecalo costs about $8–10 per tube more and can be OEM-installed or field-installed as a UL approved retrofit item.

Controlled Environment Systems, Incorporated (CESI) has since improved upon their basic Ecalo systems and evolved it into the new *Flexiwatt Lighting System*. The new Flexiwatt design approach has many advantages. With fluorescent lighting sources (lamps, ballasts and fixtures) local light levels from each Flexiwatt equipped lighting fixture can be set to just the required level of need. It then will automatically adjust up and down as required to compensate for lamp aging effects and varying daylight contribution from fenestration. The energy consumption can be reduced by as much as 80% with a corresponding 88% reduction in light output. Energy reduction from 100% use down to 69% produces a corresponding light output of from 100% to 70% and therefore performs slightly better than a one-to-one linear relationship. Below 69% energy use the reduction in power use does not correspond to a linear nearly equal light reduction but produces a slightly greater light reduction than the corresponding energy use reduction.

The circuit combines solid-state electronic switching and feedback technology with the use of reactive components including standard iron core magnetic ballasts. This unique circuitry provides control of the magnitude of the circuit current, which determines the light output, when the primary power source is an AC voltage sine wave at conventional power supply frequences. The controller can be furnished in either 120 or 277 volts and for use in Canada comes supplied for 347-volt system operation. The control method utilized is non-dissipative in the sense that losses are limited to those encountered during switching transitions and those incurred by residual dissipation in passive circuit elements.

It has distinct advantages over other competing technologies. It is cost-effective and can be employed with all conventional F40 rapid-start lamps and ballasts. It can be used with slimline and other lamps including the energy conservative types. It provides flicker-free control over a range from 100% down to 20% of rated lumen output with a corresponding reduction in energy consumption. The system does not have the dissipative losses of some competing technology products. Therefore, it has a significantly higher efficacy. In addition, it maintains a higher power factor for a given level of light output over the branch circuit control technology products.

In use, it reduces ballast heat and lamp cathode wear. Therefore, lamps and ballasts will have an extended life. The energy saved by this system will reduce a building's heat load. Accordingly, air conditioning requirement and peak demand is lowered, particularly on bright hot summer days.

This equipment can easily be retrofitted to existing fixtures or it can be added to new fixtures. It can be mounted either inside the ballast compartment or over a knock-out hole on the end or top of a fixture. For small office spaces each Flexiwatt dimmer controller can slave up to four 2-lamp, 277-volt, AC fluorescent fixtures with conventional core and coil ballasts in order to dim them or can control up to two 4-lamp, 277-volt, AC fluorescent lamp fixtures for dimming purposes.

The first energy saving feature of this system is the provision for the automatic conversion of standard ballasts and lamps into energy saving ballasts and lamps. This is accomplished by the controller reducing the amount of alternating current which is allowed to flow in the ballast circuit. Thus ballast losses are reduced and lamps are operated at reduced power. This system can operate with any magnetic ballast or fluorescent lamp as long as the sum of the rated ballast currents which are connected to it does not exceed the controllers current rating.

If energy saving lamps are being used, they can be continued and the Flexiwatt savings will still accrue. When the fixtures are relamped, a user can elect to revert to the lower cost-standards lamp for increased first cost economic advantages or continue to use the energy saving lamp.

The second energy saving feature is provided by the automatic adjustment to compensate for lumen output depreciation over operating age. Instead of having to over-light when lamps are new in order to have enough light output when the lamps are old, a Flexiwatt equipped fixture is initially adjusted down another 25% when the lamps are new. Then, as the light output begins to decline with use, the lamp's arc current is automatically increased to offset the decline.

The third energy saving feature is provided by the automatic use of daylight in lieu of fluorescent light. This is accomplished by proportionally adding or subtracting fluorescent light at each fixture location. For example, if a fixture location requires 50 fc of light and that fixture location, being close to the window, has a daylight component of 40 fc, then the fixture would only add 10 fc in order to have the necessary 50 fc. If the next fixture area (moving away from the window) has a daylight component of 30 fc, then that fixture would have to add 20 fc to have the required 50 fc. Thus, each fixture or small group of fixtures proportionally adds or subtracts fluorescent light as required to insure that the minimum set level is maintained. This per fixture daylight utilization approach is required for maximum energy conservation when daylight diminishes with distance from the window wall, or changes with the sun's azimuth.

The fourth energy saving feature is provided by the ability to adjust the light output down to as low as 12% of full output. Hence, Flexiwatt's use in nondaylight areas often saves more energy than in daylight areas. For example, energy consumption in hallways can be reduced as much as 80% with each fixture remaining at a low level. In other work and non-work interior areas such a lobbies, conference rooms, elevators, eating and lounge areas the light level can be adjusted to only the necessary level. The per fixture control approach also allows for reductions over CRT displays, files and walkways and provides a means to solve most visual acuity problems. Acceptable lighting levels range from 5 fc for hall lighting to 30 fc for conference and file areas, and 50 fc for general office areas. Some specialized work areas may require more. There are additional energy and economic savings which result from Flexiwatt's use. For example, energy savings are maximized when days are longest and the sun is brightest. This corresponds to the peak air conditioning period. If the lighting load is reduced, the buildings heat load is also reduced. Thus, the air conditioning requirement is lessened. Both the lighting and air conditioning energy savings reduce the peak electrical demand and related costs. In addition, lamp and ballast life are extended for reduced lighting maintenance costs.

A 20,000 Hz Lutron dimming system normally will run about $70.00 per lamp for a 120/208-volt, full-range dimming capability (0% to 100% dimming control). The newer 277/480-volt system, with dimming capability of from 100% to 40% costs about $30.00 per lamp.

A conventional fluorescent dimming system using dimming ballasts and solid-state automated dimming capability will cost about $25.00 per lamp. A comparable conventional non-dimming fluorescent lighting system costs in the range of from $10.00 to $14.00 per lamp.

Still another automatic energy control system out on the market within the past few years is *The Modulighting System* by Honeywell. This system also adjusts lamp output in response to both preset schedules and available natural ambient light levels. The approach starts with the first basic ingredient and that is the use of more efficient Honeywell control ballasts in lieu of standard ballasts, thereby reducing the power consumption by 15%. This allows the ballast to run much cooler (half the ballast case temperature as a standard ballast) giving longer life while producing the same light output, same spectral power distribution and the same sound rating. This cooler running not only increases the longevity of the ballast; it also marginally reduces the air conditioning load. The Honeywell AEC approach to control is when the lights are to be on or off and control each respective area by zoned controls. Controlling the lights by zones allows a perimeter area to take advantage of incoming sunlight and reduces its electric light and energy levels in direct proportion. It also allows specific areas with high intensity illumination requirements (such as critical drafting or close computer tape work) to be separately controlled and maintained than other areas with less exacting visual requirements. The system can be programmed to turn the lights on one hour before the start of the work schedule and to reduce the illumination level and energy use during lunch periods and perhaps go to reductions in lighting levels one hour after the normal workday. It would then remain at that reduced level of usage during the cleaning schedule and shut off automatically at perhaps 9 P.M. This particular modulighting system provides a strict linear relationship between light output and power consumption. If you reduce the lighting levels 20%, 40% or 60%, the energy levels will be reduced 20%, 40% or 60%. This linearity continues down to 37% of light output and then falls off slightly. A reduction of light output down to 20% will only require 27% energy consumption. Since the system is microprocessor controlled, the lighting zones on each floor can be individually programmed for maximum economy. Whenever the activity of the zone changes, different illumination levels can be set and readjusted, and different time schedules can be programmed in and provided.

Most buildings typically consume 40% of the total electric energy for lighting whereas modulighting promises a savings of 47% of lighting energy and their related costs. The effect of other disciplines due to this reduction is to reduce impact loads on the A.C. system and thus reduce the total electric energy consumption and its attendant cost by as much as 20%. This is a very flexible system that provides optimum lighting for all work zones and that easily adapts to changing requirements, while utilizing every opportunity to conserve electric energy. It has the inherent capability of shedding peak demand load in order to reduce the cost of electricity use. This system has been used extensively in Canada but is also available in the U.S. and has the necessary UL approvals for application.

ALTERNATE BALLAST SWITCHING

As an alternate approach to either conventional switching or 20,000 Hz fluorescent dimming, a 2-level Hi-Lo switching system can be used. One switch can control the ballast of the inboard tubes of a 2′ × 4′ fluorescent fixture, while the second switch controls the ballast of the outboard tubes. This provides the option of having all of the lamps on to give a high, even illumination of up to 70 fc without spotty results. Alternate ballast switching is a poor man's dimming system, but it is very effective as it permits control of lighting levels at minimal costs. Since each switch controls two lamps in each group of fixtures, lighting and power consumption can be cut 50% in any given area while still leaving adequate illumination for most tasks. A minor, but no-cost refinement to this technique is to specify that the switch closest to the door lock jamb (that is, the switch nearest at hand) controls the ballast of the inboard lamps, while the furthest switch controls the outboard lamps. Since lighting from the inner lamps is distributed with slightly higher efficiency than that from the outer lamps, the actual percentage of total fixture illumination available by operating only the most convenient switch is approximately 53% while still consuming only 50% of the energy. Since power consumption for each pair of lamps (i.e., each ballast) is identical, there is no reason not to take advantage of this incremental increase in illumination available by the simplest of means, when only half of the lamps are required for a given task.

THREE-LAMP FLUORESCENT FIXTURES

Three-lamp 2′ × 4′ fluorescent fixtures offer a Hi-Medium-Lo switching capability not possible with conventional 4-lamp fixtures. Each fixture contains two ballasts, one controlling two tubes and one controlling one tube in each fixture, thereby producing light levels of 33%, 67% or 100% of the maximum possible, such as 70 fc. This saves lighting energy by reducing light levels.

However, since a 3-lamp fixture provides only ¾ of the maximum light available from a 4-lamp fixture, ⅓ more fixtures are required to maintain a given illumination level, such as 70 fc. Since the lighting-fixture price and installation cost of 3-lamp and 4-lamp 2′ × 4s are equal, first cost of the lighting installation will be increased by 33%.

A 2-lamp fluorescent ballast uses 17.4% less energy than two 1-lamp ballasts. Use of 2-lamp ballasts is thereby not only energy conservative in itself, but it reduces the heat load introduced into the space and hence reduces the air conditioning load. For the same total number of lamps (say 1200 lamps: 300 4-lamp or 400 3-lamp fixtures), annual energy use and costs including air conditioning load will be 5.8% higher using 3-lamp fixtures.

Thus, from both a first cost and life-cycle cost approach, 3-lamp fixtures are more expensive than 4-lamp fixtures, although 3-lamp fixtures do offer a 3-level switching capability not possible with 4-lamp fixtures.

Notably, there are at least three manufacturers who produce a 3-lamp fixture that overcomes the need for a single lamp ballast and its attendant losses. Daybrites' *LFL*-system (*Lighting for Less*) has two 3-tube fluorescent fixtures connected by a flexible whip or modular wiring system connector, with up to 10 or more feet of slack cable. One lighting fixture has one 2-tube ballast controlling its outboard lamps and the other tandem lighting fixture has two ballasts. The twin ballast fixture has one ballast to control its two outboard lamps and the other ballast to control its inboard lamp and that of the other captive tandem fixture's inboard lamp. The price per the pair is nominally $130.00 and is a very functional approach at maintaining 70 fc at an incredibly low power consumption of only 1.72 watts per square foot using the KSH K19 silvertint lens. This produces a phenomenal 40.7 fc per watt. Keystone Lighting Corporation also produces tandem wired 3-lamp fixtures with 6′, 9′ or 12′ long interconnect cables by Dual-Lite or other modular wiring connector manufacturers. The modular wiring interconnected lighting fixtures come with factory supplied and installed connector cables. In addition, Lithonia lighting fixtures utilizing three 4-foot lamps are supplied tandem-wired by various lengths of Relock Flexible modular cables or whips.

HEAT-OF-LIGHT CONCEPT

The energy conservative heat-of-light concept should be used with all fluorescent and HID fixtures. This technique incorporates return air through the lighting fixtures, in lieu of using static non-air handling fixtures. By installing fluorescent lamps in dynamic air handling lighting fixtures, the lamps' efficacy is increased by about 25% more than the same lamps installed in static fixtures, thereby delivering more lumens per watt. Although HID lamp efficacy is not affected at all, both types of fixtures will realize increases in lamp and ballast life, and the heat load injected into the working space will be reduced to the minimum possible. (See Figures 6–18 and 6–19.) Furthermore, returning fixture heat directly to the ceiling plenum results in a higher temperature differential for a relatively small quantity of air, which increases the efficiency of the cooling system.

By contrast, static (non-air handling) fixtures convert over 70% of their energy consumption into heat within the working space. This requires more air changes for the space to remove the heat, as opposed to air changes to only freshen the air. The increased air flow requires larger, more costly ductwork and air handler fan motors (which consume more energy) to handle the increased ventilation requirements. The air conditioning equipment selected to serve a static lighting installation will be less efficient and thus larger in size, since extracting heat from a large air quantity having a smaller temperature differential (ΔT) is in itself difficult and energy inefficient.

In general, the heat load created by 2 kW of lighting in a building requires one additional kW of electrical energy to drive the air conditioning systems to remove that heat load.

$$\frac{1000 \text{ watts heat load}}{0.2931 \text{ watts/Btu}} \div 12{,}000 \text{ Btu/Ton} \times 1 \text{ hp/Ton} \times 1 \text{ kW/hp} = 0.284 \text{ kW chiller input}$$

$$0.284 \times 1.7 = 0.483 \text{ kW total air conditioning load required to remove 1 kW heat load.}$$

This calculation is based upon the approximate equivalence of hp to kW for large equipment by taking into account motor and compressor efficiency. The value of 1.7 is an empirical constant developed and found reliable in twenty years of experience with large building designs and takes into account peripheral mechanical equipment such as pumps, fans and air handler unit motors.

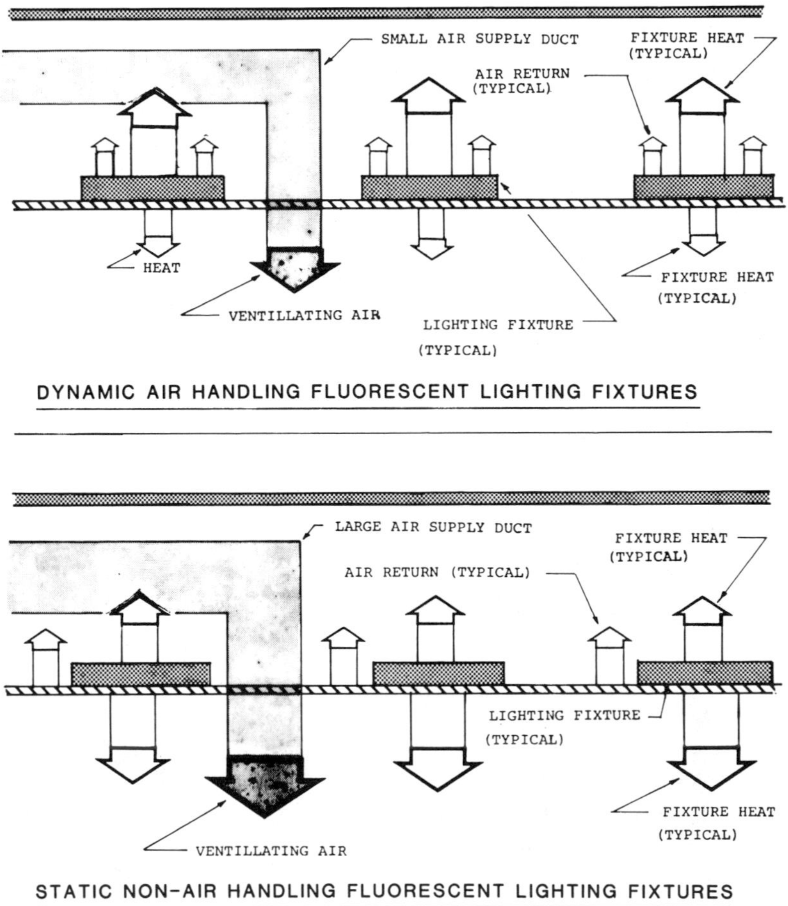

Figure 6–18. Static vs. dynamic air handling overhead lighting fixture heat analysis.

Hence, the more efficient a lighting system is, for each kW of lighting load saved, a total of 1.5 kW is actually saved. The air conditioning power consumption is defined as the electric power needed to run various motors for driving the chiller compressor and all peripheral equipment including the chilled water supply and return pumps, the cooling tower fans and the air handler unit fans.

This brief, but intense energy analysis of lighting design as part of a total building and mechanical system attempts to address the problem of design fragmentation, showing the important interdisciplinary

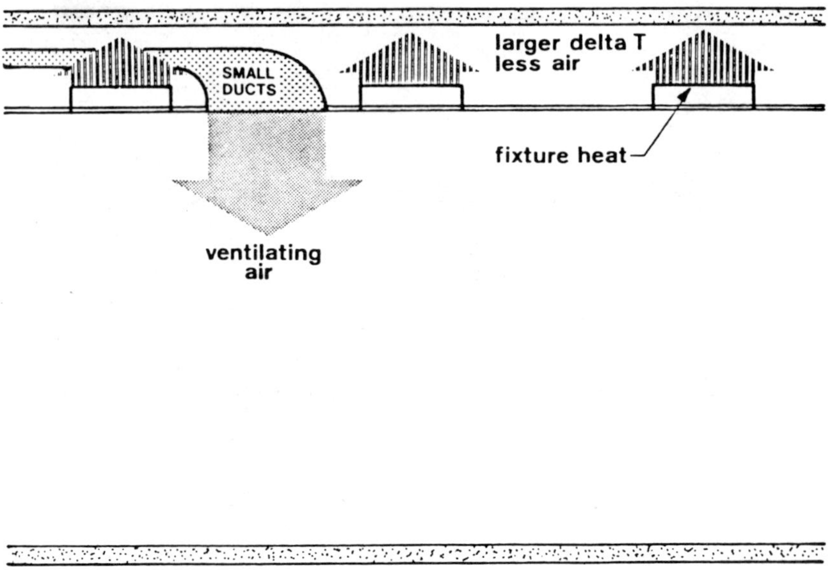

Figure 6–19. Overhead lighting heat analysis.

relationships that can not be ignored for optimum functioning and the greatest energy conservation possible.

Where dynamic air handling fluorescent fixtures are specified, pattern control dampers should be used to control the air return through the side slots, which are used to supplement the air return through the heat removal slots in the ends of the same fixture. This will avoid excessive return air movement through the side slots, which reduces the beneficial effects listed above. The heat-removal, return-air end slots that channel air through the lamp compartment will thus not be shunted or bypassed.

The lighting contractor should be directed to do the following:

- Work in conjunction with the HVAC balancing and controls specialist in adjusting the fluorescent fixtures' side slot dampers for the exact CFM required.
- Utilize, as much as possible, all air return through the lighting fixture's lamp compartment.
- Return any additional air as required through the side slot return that vents directly to the dropped ceiling cavity space.
- In no case should return air be allowed to shunt or bypass

the lamp compartment when the fixture end-slot, air-return capacity has not been exceeded.

- The heating, ventilating and air conditioning (HVAC) contractor and HVAC balancing and controls specialist should adjust the air return grille dampers so that a minimum to maximum range of 25 to 45 CFM of air will return through the lighting fixture compartment.

INTERNAL LAMP PRESSURE

Fluorescent lamps operate most efficiently at a certain internal mercury vapor pressure. If this pressure either increases due to lamp overheating (as in static, non-air handling enclosed fixtures) or decreases due to cooling (as in outdoor fixtures or dynamic air handling fixtures that supply air through the lamp compartment), light output as measured in lumens/watt will decrease. Supplying cooled, conditioned air through the lamp compartment of a louvered fixture causes a decreased temperature on the center portion of the lamp; and that portion of the lamp has a marked decrease in light output as compared to the non-cooled ends. The MV pressure throughout the lamp is decreased. Thus, the light output from the entire lamp is less, but it is markedly less in the center where chilling occurs the most. However, returning air through properly designed dynamic air handling fixtures maintains optimum lamp mercury vapor pressure and maximizes light output (and, as previously noted, also serves to extend lamp and ballast life.) The optimal ambient air temperature for rated lumen output and fluorescent lamp operation is 77°F, with the coolest spot on the bulb at about 100°F. Internal lamp MV pressure varies in direct proportion to internal lamp temperature, according to the formula:

$P_2 = \dfrac{P_1 V_1 T_2}{V_2 T_1}$ and for a constant volume fluorescent tube the

formula reduces down to: $P_2 = \dfrac{P_1 T_2}{T_1}$

$P_1 V_1 T_1$ = initial pressure, volume, temperature.
$P_2 V_2 T_2$ = new pressure, volume, temperature.

Due to lamp compartment overheating, static fixtures seldom perform according to published specifications. For example, a 4-lamp, 40-watt, recessed-grid troffer fixture without return air through the lamp compartment will experience an approximate 20% reduction in lamp light output compared to published lamp lumen ratings (as opposed to using dynamic fixtures and increasing light output by about 25% over static fixtures). Interestingly, this reduced light output also reduces energy consumption by approximately 4.3%, from the published 92 watts per 2-lamp ballast to an actual 88 watts per ballast.

This reduced real-world light output of a 4-lamp, 40-watt fixture can be taken as a baseline, as 100% of the light *actually* deliverable by the fixture. If the same fixture is then operated with 4-lamp, 34-watt lamps and a low-wattage ballast, the lower energy input will decrease the internal temperature of the lamp compartment (compared to using 4-lamp, 40-watt lamps and standard ballasts), thereby boosting light output approximately 4% above the baseline output. However, the low-wattage lamps are still operating at an elevated temperature and thus at a decreased light output (compared to published lumens), which in turn reduces the fixture energy consumption from the published 78 watts per 2-lamp ballast to an actual 74 watts per ballast.

This serves to illustrate the intricate interrelationship between internal fixture ambient temperature, light output and energy consumption. Because of the elevated internal temperature of the static fixture, conversion from standard lamps and ballasts to low wattage types saved 18.9% of the lighting energy: [(88–74) watts/74 watts] × 100 = 18.9% as opposed to the theoretical savings of 17.9%: [(92–78) watts/78 watts] × 100 = 17.9%.

At optimum mercury vapor pressure, light output per watt increases 23.5%. A 2-lamp, 40-watt, surface-mounted, wraparound lens fixture, although static, will realize a 24% increase because the two lamps produce less heat than four lamps (just as four 34-watt lamps produce less heat than four 40-watt lamps). A 2-lamp, 40-watt, industrial-type fixture, if vented, can realize a 20% lumen increase.

IN-FURNITURE, TASK-AMBIENT LIGHTING CONCEPT

Lighting consultants, interior designers, office furniture manufacturers and sales representatives are impacting the office environment by the greater use of task-ambient, in-furniture lighting systems. The function of the task element portion of the task-ambient, in-furniture lighting is to provide good quality local illumination of the task at a hoped-for lower energy expenditure. A function analysis of the in-furniture, indirect-ambient lighting discloses that its purpose is to reduce high levels of lighting contrast between brightly lighted task surfaces and darker unlit surface areas such as ceilings. These new furniture lighting systems have a substantial impact on building structures and their related disciplines in how they do or do not integrate into the system as a whole. In-furniture lighting systems are still another new technology to be considered along with other time honored trades, e.g., electrical, mechanical, etc. In-furniture, task-ambient lighting has been touted in the past several years as the best thing since sliced bread. It does have several problems not generally acknowledged or understood. These outweigh its advantages: aesthetics and the new look.

In-furniture, task-ambient lighting technology and equipment are used primarily in office buildings. This analysis of lighting design as a part of total building electrical and mechanical systems will address the problem of design fragmentation, showing the important interdisciplinary relationships which must be considered to obtain an optimum design. As a general framework within which to conduct the analysis, we will assume that an energy efficient lighting design should employ light sources (lamps) of the highest efficacy so as to reduce the number of fixtures and hence reduce costs. That lighting, and as much of the building load as possible, should be operated on 277 volts. Due to its higher load carrying capacity, a 277/480-volt distribution system can carry 231% as much energy as a 120/208-volt system.

Lighting by furniture mounted sources has been considered in recent years as an alternative to general overhead lighting. In general, task lighting normally reduces overall lighting levels throughout an installation; but energy use in watts per square foot may be greater than that with general overhead HID or fluorescent lighting, depending upon the number of task lighting fixtures used. (See Figures 6–20, 6–21, 6–22 and 6–23.)

We will now develop a functional analysis of the combined effect of both task and ambient lighting in order to analyze the types and number of light sources required. Based on considerations of reduced first cost and life-cycle costs, a cost comparison can be made based on equipment efficiency, usable life, material and labor cost of replacement of equipment. In addition, an energy load profile in watts per

PROHIBITIONS ON THE USE OF 277 VOLTS FOR TASK LIGHTING:
1981 NATIONAL ELECTRICAL CODE

210-6. MAXIMUM VOLTAGE

(a) VOLTAGE TO GROUND. BRANCH CIRCUITS SUPPLYING LAMPHOLDERS, FIXTURES, OR STANDARD RECEPTACLES RATED 15 AMPERES OR LESS SHALL NOT EXCEED 150 VOLTS TO GROUND.

EXCEPTION NO. 2: THE VOLTAGE SHALL NOT EXCEED 300 VOLTS TO GROUND ON BRANCH CIRCUITS IN INDUSTRIAL ESTABLISHMENTS, STORES, HEALTH CARE FACILITIES, OFFICE BUILDINGS, SCHOOLS, OR PUBLIC AND COMMERCIAL AREAS OF OTHER BUILDINGS, SUCH AS HOTELS OR TRANSPORTATION TERMINALS, WHERE ALL OF THE FOLLOWING CONDITIONS ARE MET:

a. THE BRANCH CIRCUITS SUPPLY ONLY THE BALLASTS FOR ELECTRIC-DISCHARGE LAMPS MOUNTED IN PERMANENTLY INSTALLED FIXTURES.
b. THE FIXTURES DO NOT HAVE MANUAL SWITCH CONTROL AS AN INTEGRAL PART OF THE FIXTURE.
c. ELECTRIC-DISCHARGE LAMPHOLDERS OF THE SCREW-SHELL TYPE ARE MOUNTED NOT LESS THAN 8 FEET ABOVE THE FLOOR.

COMPARISON OF TASK LIGHTING LAMPS AND BALLASTS

LAMP TYPE	TOTAL LUMENS & EFFICACY IN LUMENS/WATT	LAMP COST	LAMP LIFE	120 VOLT 2 LAMP BALLAST COST	BALLAST LIFE
FLUORESCENT 40 WATT RAPID START (ALSO STANDARD OVERHEAD LIGHTING LAMP)	3150 78.75L/W	$2.30	20,000 HOUR (UP TO 37,700 WITH A.E.C.)	$12.63	STATIC 13-15 YEARS DYNAMIC 15-18 YEARS
FLUORESCENT 15 WATT TRIGGER START (DESK LAMP)	780 52L/W	$2.95	7,500	$16.14	STATIC 6-8 YEARS
FLUORESCENT 20 WATT TRIGGER START	1260 63L/W	$2.75	9,000	$16.14	STATIC 6-8 YEARS
FLUORESCENT 30 WATT RAPID START	2200 73.3L/W	$4.55	7,500	$13.81	STATIC 6-8 YEARS
INCANDESCENT 100 WATT	1740 17.4L/W	$ 69	750	--	--

Figure 6–20. Task analysis.

square foot (W/ft^2) can be developed and modeled based upon current and conventional task spacings and ambient lighting fixture spacings. But before we delve into those analyses we must first ascertain the minimum levels of quality low-glare illumination that will be acceptable, based on first-cost tradeoffs. To evaluate quality lighting and how it can be achieved, the following concepts must be understood.

FLUORESCENT LAMP WATTAGE	NUMBER OF LAMPS PER BALLAST	COST OF 120 VOLT BALLAST	COST OF 277 VOLT BALLAST	LIFESPAN OF BALLAST
15 WATT TRIGGER START	1	$16.14	$17.10	STATIC FIXTURE 6-8 YEARS
	2	$16.14	$17.10	DYNAMIC FIXTURE 8-10 YEARS
20 WATT TRIGGER START	1	$16.14	$17.10	STATIC FIXTURE 6-8 YEARS
	2	$16.14	$17.10	DYNAMIC FIXTURE 8-10 YEARS
30 WATT RAPID START	1	$13.81	$14.55	STATIC FIXTURE 6-8 YEARS
	2	$13.81	$15.00	DYNAMIC FIXTURE 8-10 YEARS
40 WATT RAPID START	1	$12.05	$12.63	STATIC FIXTURE 13-15 YEARS
	2	$12.63	$13.50	DYNAMIC FIXTURE 15-18 YEARS

Figure 6–21. Comparing lamp ballast. (Cost and life expectancy of various wattage one and two lamp ballasts).

Zones of direct and indirect glare come from lighting fixtures located in front of the worker. The offending zone for locating lighting fixtures is in an area in front of the worker at an angle that will allow indirect glare to produce veiling reflections. A veiling reflection can be described as casting a light source image or veil upon the task that reduces the contrast of the printed work or diagram to the background material it is printed on. This condition increases the difficulty of reading, increases the percentage of errors, reduces worker efficiency while increasing eye strain and worker fatigue. Direct glare from a luminaire (lamp and fixture) mounted directly in front of the worker primarily induces eye strain as well as worker fatigue leading to lower productivity and a higher percentage of errors. This condition can be mitigated by using various shielding media such as lenses and louvers to shield the light source and to control the direction of the light. Batwing lenses can be installed in the lighting fixture so as to distribute the light to both the right and left of the worker in lieu of straight ahead, directly and indirectly into his eyes. However, workers being impermanent objects that are not nailed down to one spot will tend to reorient themselves at times to one side of the desk or the

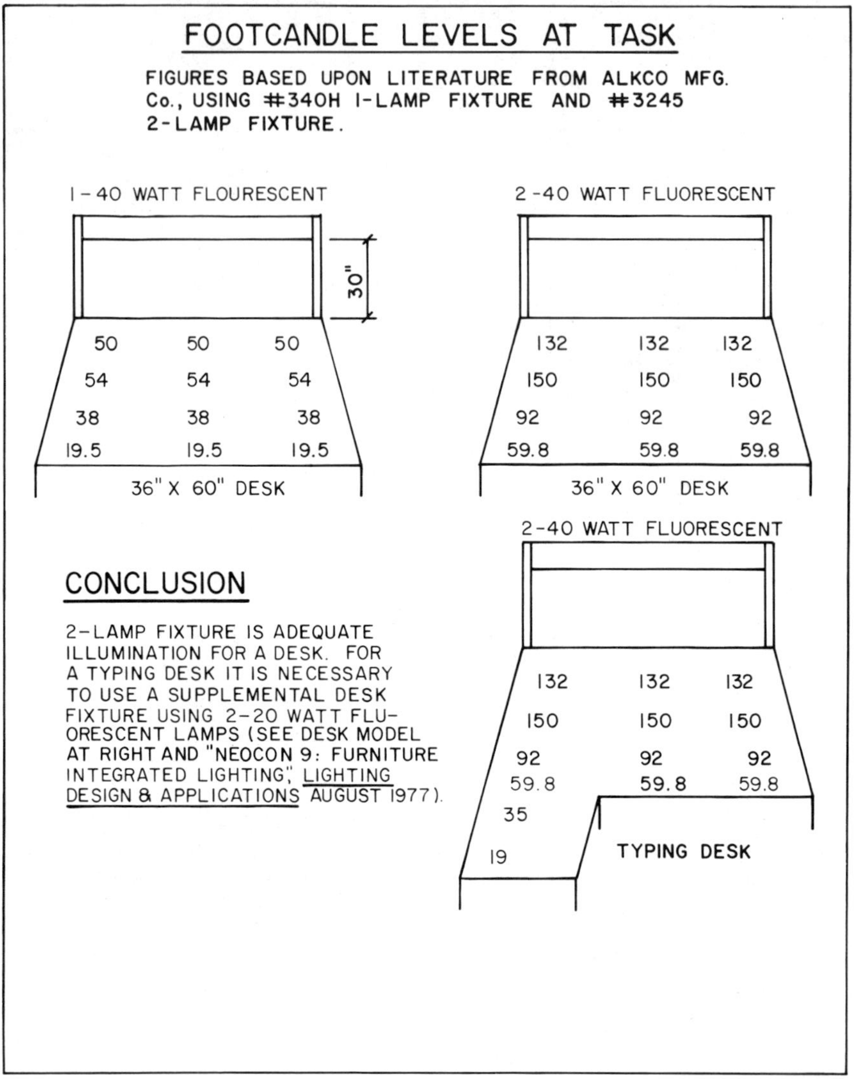

Figure 6–22. Task lighting

other. At such times the redistributed light from the batwing lens causes the same deleterious visual environment previously described for a conventionally shielded fixture installed directly in front of the worker.

From the standpoint of good quality, low-glare illumination on an office desk or study carrel, the desk or study carrel should ideally have

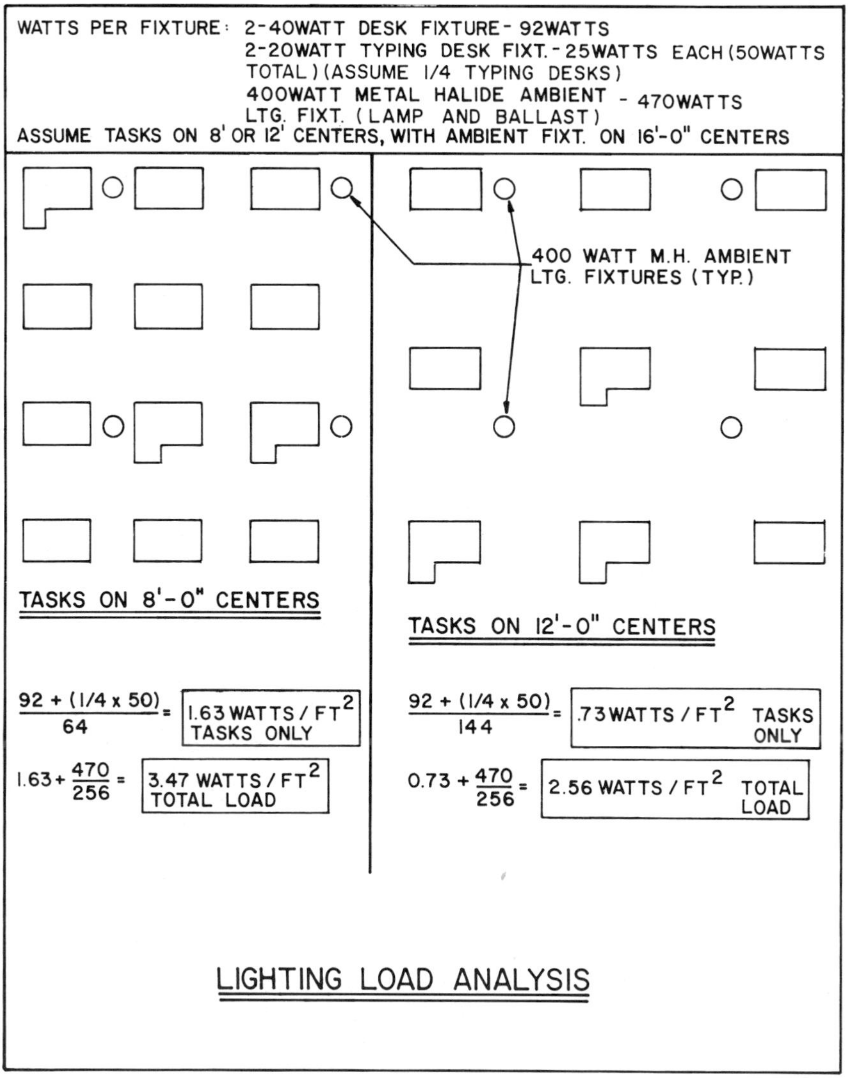

Figure 6–23. Task and indirect ambient lighting.

the lighting fixture to either or both sides of the viewer, that is, to his or her immediate left- and right-hand sides. With a good quality shielding medium the worker receives no direct glare into his eyes. The angle of incidence of the light from the two sources precludes light bouncing off the task material and reflecting into the worker's eyes; thus, disability glare from veiling reflections is virtually eliminated.

A standard 36″ wide by 60″ long desk used for conventional office paperwork requires the use of either two single-lamp, 40-watt fluorescent fixtures on either side of the worker or just one 2-tube, 40-watt lamp fixture mounted directly in front of him. (See Figure 6–22.) Calculations were based upon literature from Alko Manufacturing Co. using a #340HO 1-lamp fixture and #3245 2-lamp fixture data in addition to extensive in-the-field investigations and search of all available literature on this subject matter. For typing desks the quick dropoff of illumination due to runoff at the extended typing location requires a supplemental desk fixture using two 20-watt fluorescent lamps. The calculations shown on the desk model in the bottom right-hand side of Figure 6–22 confirms the need for this supplemental lighting. In addition, a search of the literature also confirms this in the *NEOCON 9* (Furniture integrated lighting, *Lighting Design and Application,* August 1977).

At this stage of our modeling analysis we must make an economic analysis and trade-off of the advantages and disadvantages based upon first cost of installation and life-cycle cost due to additional energy expenditure. As can be appreciated from the data in Figures 6–10(A), 6–10(B) and 6–20, we have attempted to use lamps having the highest efficacies, longest lives and lowest costs. In addition, the peripheral or ancillary equipment such as ballasts were evaluated in conjunction with their respective lamps in terms of first cost, life-cycle costs and life expectancy. (See Figures 16–13, 16–20 and 16–21.) Regardless of their other characteristics, fluorescent lamp ballasts should be of the 2-lamp type, and the use of single-lamp and single-lamp ballast fixtures should be minimized as much as possible if not eliminated. A single 40-watt lamp plus ballast consumes 52 to 54 watts (depending upon manufacturer) and two 40-watt lamps plus ballast consume 92 to 94 watts. Thus, two single-lamp fixtures can consume as much as 108 watts or 17.4% more energy than a 2-lamp single ballast fixture at only 92 watts.

A single-lamp, 40-watt, RS fluorescent tube fixture not only costs upwards of 70% of the cost of a 2-tube ballasted fixture, it also incurs the same cost of installation twice. Since twice as many single tube fixtures as 2-tube fixtures are needed, the material cost of installation may be 40% higher in addition to a 100% higher installation cost. Thus, one 2-tube, 40-watt, fluorescent fixture should be used per conventional worker desk in lieu of two single-tube, 40-watt fluorescent fixtures for the purpose of minimizing first cost as opposed to increasing the quality of the lighting installation.

We will assume that four desks utilize one 2-tube fixture requiring 92 watts each. The fourth desk is assumed to be a typing desk requiring an additional supplemental lighting fixture using one 2-tube, 20-watt lamp fixture consuming 50 watts. Although we have somewhat compromised the quality of the lighting installation by reducing the number of fixtures, we do not have an inordinate number of different fixture types, lenses and lamps. If we had not done this we would have ended up with a logistics problem of stocking all these various commodities. As it is, in lieu of a reasonably large quantity of only a few major commodities with task lighting by overhead ambient lighting fixtures, we are faced with fairly large stockpiles of many different types of major commodities for in-furniture, task-ambient systems. One of the results of this is that the per unit cost of the items is higher since discounts are generally extended only to large bulk purchasing. As a result, 100 units of 100 different items of equipment will cost more per unit for those 10,000 items than say 400 units of only 10 different items or 4,000 pieces of equipment.

Secondly and probably the most important aspect of this problem is that uniformity of appearance is sacrificed by incorrect replacement of parts due to inept maintenance or confused maintenance personnel. Desired results are often lost either due to lack of prompt proper replacement of lamps or lenses, or incorrect replacement parts being installed. Incorrect replacement might consist of using a 400-watt MV lamp in a 400-watt, MH, indirect-ambient, floor-lamp fixture, causing loss of upwards of 50% of the ambient lighting. It could also consist of replacing a desk task lighting fixture with a non-divergent batwing lens, or leaving the lens out altogether thus producing discomfort glare, such as direct and indirect glare which in turn produces veiling reflections and loss of contrast. As a result, in no time at all this ultramodern and sleek task-ambient, in-furniture system is rendered shabby, poorly maintained, poor in lighting quality and high in operating maintenance costs.

In addition, we previously defined the need of the ambient lighting fixture as required to reduce contrast by producing fill illumination and to reduce worker fatigue due to eye strain. This form of eye strain is produced by the eye having to make rapid adjustments while scanning from brighter objects illuminated by higher levels of illumination to darker objects at lower levels of illumination. One deleterious aspect of in-furniture, indirect-ambient lighting is that it produces pools of hot spot lighting immediately above the desk, table or pylon mounted ambient light source. This is especially true for HID lamps such as MV and MH. The higher wattage HID lamps are more efficient (greater efficacy) and have a greater life span in the larger sizes, especially the MH lamps. As a result, it is not uncommon to see desk, table or pylon mounted 400-watt, MH, indirect-ambient lighting systems in use. In order to properly and efficiently distribute this indirect-ambient light, a hard, somewhat glossy reflective ceiling surface is often required. This somewhat alleviates the high hot spot ceiling contrast and distributes the light a little more efficiently by having a higher reflectance factor. However, hard flat reflective surfaces do not effectively attenuate high ambient background office sounds. As a result, the ceiling tiles' sound transmission coefficient may increase from 20 to 50 and sound transmission reduction may be minimized or lost since the ceiling-tile noise-reduction coefficient is increased rather than decreased. The frustrating aspect of this is that you cannot have the one benefit without the deleterious effect of the other, and vice versa.

If indirect-ambient lighting fixtures use 400-watt MH or MV lamps, they consume a total of 470 watts (counting ballast loss). If consistently located on 16-foot centers or one per 256 square feet they will contribute 1.836 watts/ft^2 of energy consumed (470 watts/246 ft^2). If the task furniture is located on 8-foot centers, or one desk per 64 square feet, at 92 watts each, plus 50 watts additional for every fourth desk, that contributes 1.63 watts/ft^2 of energy consumed. (See Figure 6–23.) Thus, the total task-ambient energy load is 1.836 watts/ft^2 + 1.63 watts/ft^2 for a total consumption of 3.47 watts/ft^2. If the tasks are spaced farther apart on say 12-foot centers, or one per 144 square feet, while retaining the same spacing for the ambient lighting fixtures, the load per square foot can be seen to drop to only 2.56 W/ft^2. Should the tasks be spaced on 16-foot centers (256 square feet per task) with the same ambient-lighting fixture spacing, the load becomes lower, but still a high energy consuming level of 2.24 W/ft^2. Task spacings increased to 20 feet on center (400 ft^2) produce 2.097 W/ft^2.

Thus it can be appreciated that spacing of the tasks has a lot to do with the energy load profile in terms of energy consumed in watts per square foot. It is just not possible in the context of task-ambient lighting to address all installations as the same by saying that regardless of the task-spacing the energy consumption remains the same. Nor it is realistic to claim that in-furniture, task-ambient lighting can produce illumination levels comparable with task lighting from overhead ambient lighting fixtures at levels of energy consumption in the under 2 W/ft^2 range.

Task lighting has aesthetic qualities which may be desirable in some areas of a building and, in addition, should be used to supplement general overhead lighting where high local footcandle levels of illumination (of up to 200 fc) are required for critical tasks such as drafting or close computer tape work. A combination of task and overhead lighting is the most economical method of realizing high local levels of illumination. However, task-ambient lighting furniture and equipment without overhead lighting is not energy efficient. For purposes of energy conservation should not constitute a major part of a total lighting installation.

A primary limitation of task- or indirect-ambient lighting as compared to overhead lighting, is the requirement that it be served with 120 volts to comply with the National Electrical Code. It is desirable from a power distribution standpoint to serve as much as possible of the lighting and other building loads at 277/480 volts. Large numbers of 120 V, in-furniture lighting fixtures will increase the number of branch circuits required, necessitating more panelboards and larger step-down transformers, feeders and distribution equipment. (See Figure 6–24.) To obtain a 57% reduction in current, a service voltage of 277/480 volts, 3-phase, 4-wire should be used for all balanced 3-phase loads such as motors, electrical resistive heating and for all overhead lighting systems, as opposed to using the more commonly used, 120/208-volt, 3-phase, 4-wire system. This will result in a more than 60% reduction in copper conductors because of the non-linearity in current carrying capacity for larger conductors (ampacity does not increase equally for equal increments of conductor size). Logically, a reduction of 60% or more in size and quantity of raceways follows. This one design aspect of the total overall project alone can save upwards of 20% of the total electrical project cost.

There are additional factors which should be considered in the use of task- or indirect-ambient lighting. The 1981 NEC states that 180 voltamps at unity power factor or watts shall be allowed for every receptacle. Also, NEC Article 384–16(c) requires that "the total load

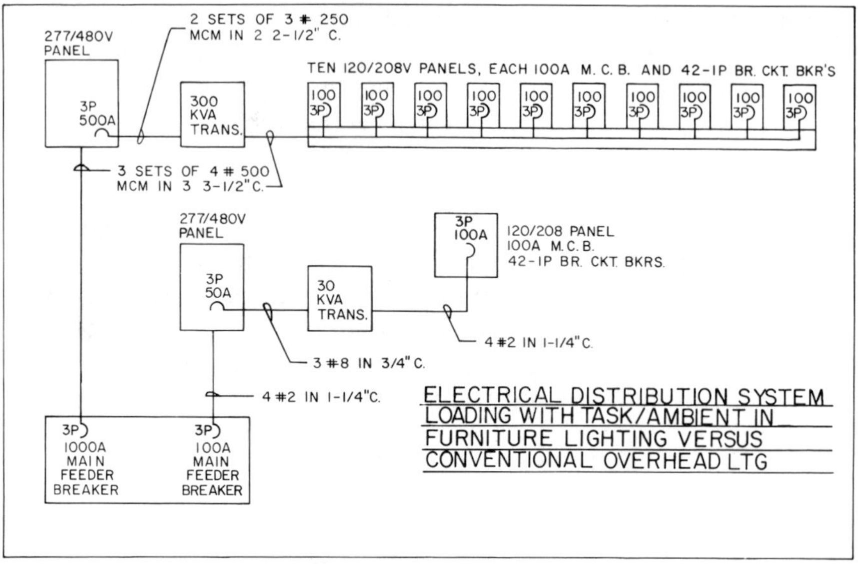

Figure 6–24. Load analysis of various systems.

on any overcurrent device located in a panelboard shall not exceed 80 percent of its rating where in normal operation the load will continue for 3 hours or more," except "where the assembly including the overcurrent device is approved for continuous duty at 100 percent of its rating." Since a conventional molded case branch circuit breaker is allowed to carry only 80% of its load continuously, a 20-amp, 120-volt branch circuit breaker is capable of carrying 16 amps continuously: 1920 watts (120 volts × 16 amps). Ten receptacles equals an 1800-watt load (180 voltamps ×10 receptacles). Thus, ten receptacles is the maximum load a 20-amp, 120-volt branch circuit can carry (See Figure 6–25.)

Since integrally switched fluorescent lamps and screw-shell-base lampholders for HID lamp fixtures mounted below 8 feet must be served at a voltage not to exceed 150 volts to ground according to NEC Articles 210–7(a)(d) and 210–6(a)(c), these task- or indirect-ambient lighting fixtures must be fed from 120-volt convenience receptacle circuits. It then becomes necessary to consider the effect of reflected ambient lighting from in-furniture, 400-watt, mercury vapor or metal halide HID sources on the 120-volt circuitry. A 400-watt HID fixture with ballast consumes 470 watts; so a 120-volt, 20-amp branch circuit breaker can carry only four of these (1880 watts). Since these fixtures

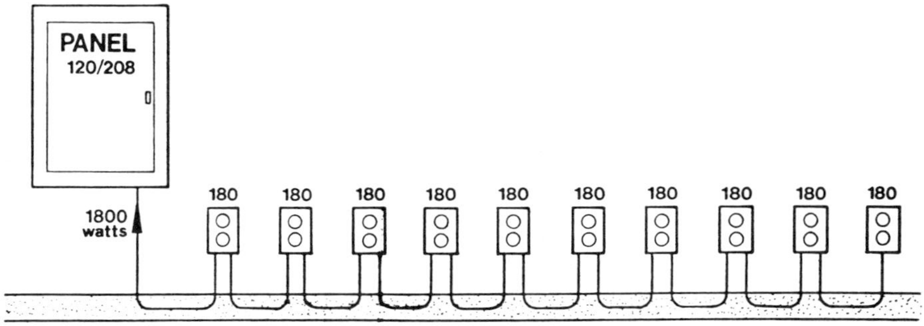

Figure 6–25. Circuitry load.

will be connected one per receptacle, then only four receptacles per circuit can be allowed due to the higher load. With only four receptacles (without additional loads) per 20-amp circuit rather than the ten allowed by code, 250% more branch circuits, 250% more branch circuit breakers and 250% more panelboards will result. (See Figure 6–26.) If, however, two lighting fixtures are connected to each duplex convenience receptacle, then only two receptacles can be connected per each branch circuit. Since only two receptacles can be connected on each 20-amp, 120-volt circuit as per code (in lieu of ten), a 500% increase will be required for branch circuits, branch circuit breakers and panelboards. For the same lighting load the same number of branch circuits are required, i.e., whether there is one or two lighting fixtures per receptacle.

The Washington, D.C. electrical code allows that conventional receptacle loads can have a demand factor placed on the load for receptacles since they are not used more than 10% to 30% of the time. This reduction in the demand load permits a reduction of the main lugs or main circuit-breaker rating of the various panel-boards and is reflected in smaller panelboard feeders, smaller 480-volt to 120/208-volt step-down transformers, transformer feeders and feeder breakers, as well as smaller main distribution equipment. However, with task-ambient lighting fed from convenience receptacle branch circuits, none of these reductions can be applied. Thus, rather than increasing the overall size of the 120/208-volt distribution system 250% by requiring it to serve lighting, the actual increase may be as high as 833%, where 0.3 is the demand factor applied to the receptacle circuits. The demand factor cannot be used when serving a constant lighting load (250%/0.3 = 833%).

It should be noted that while the NEC does not allow these task

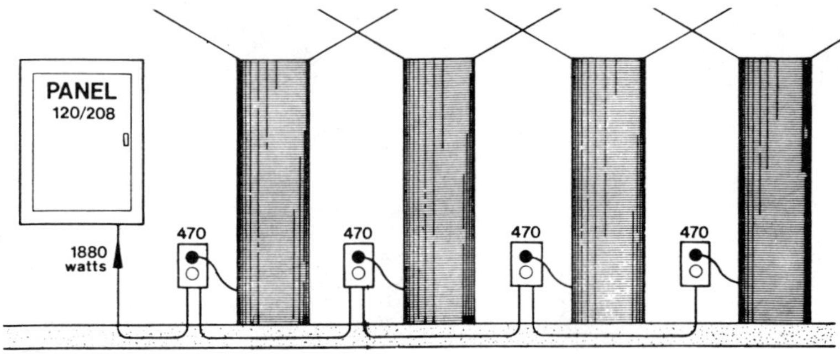

Figure 6–26. Indirect ambient lighting.

lighting loads to be served at a voltage higher than 150 volts to ground, it does not preclude the use of 208 volts (line to line) which is only 120 volts above ground. However, 208 volts for lighting circuits is neither a good use of materials nor optimum engineering practice, since it increases both first and life-cycle costs. 208 volts is vectorially the combination of two 120-volt vectors, expressed mathematically as: $120\text{ V}\sqrt{3} = 120\text{ V}(1.732) = 208\text{ V}$.

Thus, for a given circuit breaker trip setting and a 208-volt circuit voltage, its wattage has been increased by only 73%. However, a 208-volt, single-phase circuit breaker is a 2-pole device (a 100% increase in material). Further, two 120-volt, 20-amp circuits can transport 3840 watts ($2 \times 120\text{ V} \times 20\text{ A} \times 0.8$), while one 208-volt, 20-amp circuit can transport only 3328 watts ($208\text{ V} \times 20\text{ A} \times 0.8$). Thus, two 120-volt circuits can deliver 15.4% more energy than a single 208-volt circuit, or put another way, one 208-volt circuit can deliver only 73.2% more energy than a single 120-volt circuit. Of course, the single 208-volt circuit requires only two phase wires versus two phase wires plus a neutral (three conductors) for two 120-volt circuits, but this is a negligible saving compared to the cost of more and larger panelboards and more branch circuit breaker poles.

Thus, 208-volt circuits not only increase the first cost of the electrical distribution system but also increase the life-cycle costs due to the greater maintenance required by the additional, larger equipment. More lower voltage equipment for the same kilowatt load will result in larger voltage drops and increased energy wastage due to the higher current densities incurred by the lower voltage system. The increased number and sizes of the step-down transformers used also introduces a larger energy wastage because of the transformers' core and coil losses, which can be quite high in smaller sizes such as a 15 kVA unit. This, in effect, is double jeopardy; to serve a given load, the transformer loss is now incurred twice, once in the main (primary voltage to 277/480 volt) transformer at approximately 1.85% loss and again at up to 4.76% in the (480 volt to 120/208 volt) step-down transformer. Therefore, although a task lighting design may under some design conditions reduce the raw watts per square foot for lighting only, when viewed as part of the total electrical system it tends to increase the total watts consumed for the whole system and increases both its first cost and the life-cycle costs of the project.

Task-ambient lighting not only produces its light in the working space but converts all of its power consumption into heat within the working space. More air changes are then required to remove heat from the space than would be required merely to freshen the air—meaning larger, more costly ductwork and larger air handler fan motors (which consume more energy) to handle the increased ventilation requirements (See Figure 6–27.) The air conditioning equipment to serve a task-ambient lighting installation will also be less efficient and thus must be larger, since extracting a small temperature differential (ΔT) from a larger air quantity is in itself difficult and energy

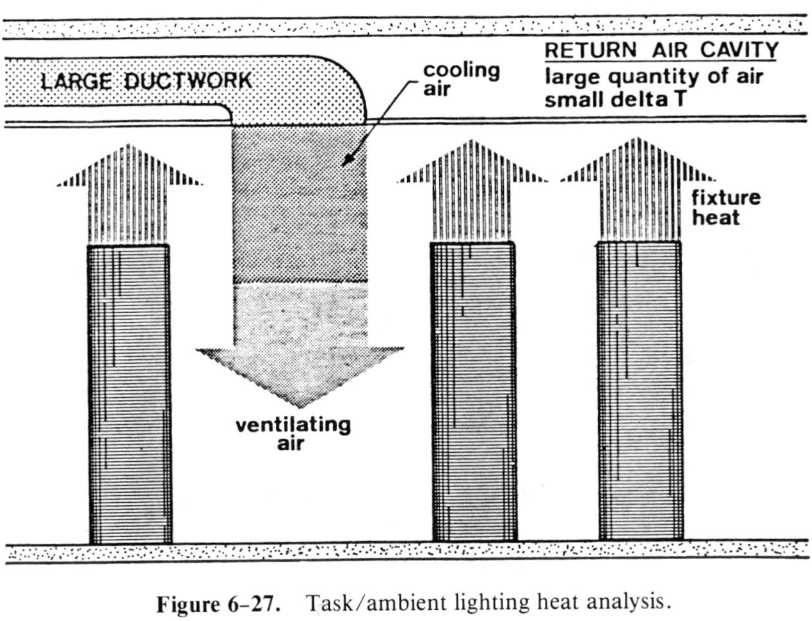

Figure 6–27. Task/ambient lighting heat analysis.

inefficient. If conventional overhead lighting incorporating the heat of light principle is used, then the plenum return air temperature difference (ΔT) is larger and easier to extract. More efficient and smaller air conditioning equipment can be used. Since with conventional overhead lighting more of the heat will be in the ceiling cavity space, less ventilating air is required for the working space, with attendant reductions in the size of ductwork and air handler motors. (See Figures 6–18 and 6–19.)

One manufacturer that appears to have overcome the deficiencies of most task-ambient in-furniture lighting systems, as previously enumerated, is Elliptipar. Theirs is a unique and different approach, but like all systems, the concept must be approached prudently and the system applied with care. This care must be taken in order to maximize the benefits while minimizing any possible difficulties or deficiencies.

The Elliptipar, Inc.'s Lite-A-Part is a true task-ambient lighting instrument; i.e., both task and ambient light can be provided by one fixture, as well as from one lamp. Underlying the instrument's design is a distinctive lighting philosophy evolved by its designer, Sylvan R. Shemitz.

Ambient lighting should be soft and diffused, filling the space with glare-free illumination that balances the brightness between a worker's immediate surroundings and the overall environment. Ambient lighting expresses the contours and dimensions of the work space and enriches its details. It must also be sufficient for carrying out duties away from the desk safely and efficiently. Mr. Shemitz's belief is that indirect, fluorescent light best fulfills these needs, for a linear source can distribute light over a wider surface than a point source. Many choices are available of both output and color.

Task lighting should provide light levels appropriate to the task for visual comfort and productivity. Brightness values throughout the task area should be relatively even to eliminate the need for frequent eye adaptation. It should be free of direct glare throughout the range of probable seated eye heights. It should minimize veiling reflections and be positioned to eliminate furniture, body and hand shadows.

Still other requirements include applicability to the widest possible range of furniture systems (while still maintaining instrument standardization); energy efficiency; compatibility with a range of ceiling heights; ease of installation; simple to lay out; rugged enough construction to withstand handling during furniture rearrangement; use with a wide range of standard, inexpensive lamps; and quiet, heat-free operation.

The Lite-A-Part emerged as a response to these requirements. The fixture consists of two structurally ribbed, extruded aluminum side panels. An aluminum endplate completes the rectangular enclosure. Within is room for a complex, variable system of optics, a ballast compartment and space for optional features such as convenience outlets and switching. The basic models are 6″, 9″ or 13″ wide and 4½″ high. Length can be varied to satisfy need (up to 12 feet). The unit is generally mounted within a partitioned furniture system.

Ambient light is provided by the upper half of a single fluorescent lamp surrounded by a series of reflectors that throw light upward and outward. A grid or louver composed of parabolic sections sits above the lamp in a gasketed track (to prevent vibration). This is a low brightness shield to prevent glare with the fixture mounted below standing eye height. Its 114° distribution projects light widely.

Task lighting is provided by the lower portion of the same fluorescent lamp. A segmented asymmetric reflector projects its light downward and outward through a batwing lens that reflects light at 35° to both sides. The batwing serves a dual function; light is angled so as to be nonparallel with most sight lines, substantially reducing veiling reflection; and the side partition is illuminated. Opposite the batwing lens is a splay reflector that lights the back partition. This unique combination of optical components, if utilized with appropriately reflectant surfaces, provides balanced brightnesses for the worker's entire micro-environment.

The unit's downward distribution can be varied, during manufacture, to adapt the light distribution to different partition heights and desk depths. For shallow desks and high partitions, the lamp/reflector relationship lifts the source out of the glare zone and concentrates the light within the shallow work zone. For wide desks or lower partitions, it is altered to project the peak candlepower in a shallow trajectory aimed at the front edge of the surface. This maximizes the use of available lumens.

An optional roller track enables the instrument to be moved forward or backward in the task area. This feature enables the illumination of runoff areas. A useful application is to light VDT equipped stations with precision; light can be directed to all points within the VDT equipped station—even the keyboard—without creating bright images on the screen, the bane of the operator's existence and the chief obstacle to improving his productivity. There is also an optional enhanced cutoff shutter. This device, mounted on the front panel, may be raised or lowered to prevent light from being reflected onto the screen from the operator's shirt front or a nearby wall opposite the screen.

The lighting of drafting stations is another excellent application of the roller track. As the board is tilted, the unit can be slid toward the worker; or if the worker wishes to do extensive work at the runoff, sliding it closer gives good illumination free of veiling reflections.

Energy efficient results from directing the energy from the lower half of the fluorescent lamp to the task and utilizing the energy from the top of the lamp for indirect lighting. Typically, Lite-A-Part installations use about one watt per square foot using conventional ballasts and .75 watt per square foot using "energy saving" types. This produces about 50–55 FC on the desk surface and 12–20 FC in the environment. (The use of high output lamps raises this to about 70 FC and 17–20 FC, respectively.) Lighting of the environment aids task-ambient lighting. Wall-washing surfaces such as core walls are a particularly effective approach. It is an antidote to contrast between often viewed surfaces.

ENERGY ACTIVATION SWITCHING EQUIPMENT (E.A.S.E.)

E.A.S.E. is an electric switching system that is activated by the movement of a person. It is designed to control room lighting by automatically turning off fixtures when an area is not in use. The system is based on the potential disturbance of a field of inaudible ultrasonic sound waves which fill the monitored area. Motion of a person in the area is detected by the system's sensor and this motion causes switching to take place, thereby turning the lights off after an adjustable delay of from 90 to 360 seconds.

The E.A.S.E. system consists of an oscillator(s), air couplers and sensors, switch modules, relays and power supply. Only the air couplers and sensors are visible as small discs mounted flush in the ceiling. The air coupler permeates the area covered with inaudible airborne ultrasonics. The movement or presence of a person is detected by the sensor. Hidden parts of the system receive and relay this information, causing the electrical system to respond. E.A.S.E. is sensitive to the most minute movements; the breathing of a perfectly stationary person is sufficient to maintain room lighting in a normal "on" condition.

The system is compatible with fluorescent lighting which restrikes instantly, and with HPS and High CRI HPS which restrike in one minute after extinguishment and produce approximately 80% intensity after 7–10 minutes, and restrike immediately after extinguishment if dimmed down to 48% or less of energy use and light output, reaching full intensity within two to three minutes, respectively. E.A.S.E. is not suitable for use with MH or SMH lighting, which requires from 10 to 15 minutes to restart after extinguishing and several minutes more to reach full intensity.

With the E.A.S.E. system, lighting cannot be accidentally left "on." It is particularly useful for lighting control in areas used intermittently, such as conference rooms, storage rooms and service spaces generally, where lights left burning needlessly would often go unnoticed for long periods of time.

A payback period of 5 years or less can be expected for large installations.

LIGHTING AND ENERGY SAVER SYSTEM (L.E.S.S.)

L.E.S.S. was developed by Novitas, Inc., in 1977 and is a system similar to E.A.S.E., controlling lighting through ultrasonic sensing of personal movement. It is designed to work in conjunction with an automatic dimming system such as an AEC or Hi-Lume system, as well as turning fixtures "on" and "off." Like E.A.S.E., L.E.S.S. incorporates an adjustable time delay (30 seconds to 15 minutes). This system has since been replaced by *Light-O-Matic,* introduced by Novitas, Inc., in 1978. The Light-O-Matic automatic lighting fixture controller by Novitas, Inc., is an adjustable ultrasonic device that can sense the presence of a person. It turns lights on immediately upon a person's entering a room and extinguishes the lighting fixtures after the room has been vacated up to five or more minutes. Light-O-Matic has been somewhat more successful than its competition in that it is simpler and less expensive. It is totally modular and can be used in small individual office spaces of only 100 square feet, or in large open architectural office spaces of up to 25,000 square feet. There are several modules that cover a range or applications:

- A recessed unit with up to 150 ft^2 coverage.
- Another recessed unit that will extend the coverage an additional 200 ft^2 and is auxiliary to any of the other sensors in the system.
- A suspended unit that covers rooms up to 500 ft^2.
- Another suspended unit that extends coverage an additional 500 ft^2 and is also auxiliary to the other sensors in the same system.
- A recessed residential unit with integral photocell that prohibits inadvertently turning lights on in the daytime when there is adequate illumination.

The following energy and attendant energy cost savings are attributable to this product by its manufacturer: 33% to 55% for individual offices, 20% to 33% for large open offices, 33% to 65% for conference rooms, 35% to 55% for classrooms and 35% to 85% for storage areas.

INFRACON

Another new automatic lighting controller came out in 1982 and is actuated by heat in the 7 to 14 micrometer (μM) range rather than by inaudible ultrasonic sound wave sensing. Infracon is an electronically controlled passive infrared sensor and is a product of Tishman and United Technologies. They provide many other electronic control and energy management systems that reduce energy costs and improve building operations. The sensor detects and responds to changes in radiated heat within a room caused by the presence or movement of an individual. The Infracon system consists of a controller that is designed for use with either one or two sensors. The sensor assembly is a high impact polystyrene cylinder that houses an infrared sensor and low voltage control equipment in conjunction with accessories such as an optical filter and an aperture ring.

The sensor will detect any change in heat radiated by a person entering the room and will actuate the lighting system. The lighting will be maintained in an operating mode by the individual's body heat and movements as long as he remains in the room. A delay period of approximately 12 minutes after departure of the occupant causes the system to extinguish the lighting. This built-in time-delay feature prevents annoying switching of the lights due to an individual leaving the room momentarily or remaining motionless for short periods of time. One sensor in conjunction with a basic control system will provide coverage for a room approximately 14′ × 14′ or nearly 200 square feet. The maximum number of sensors per each controller is two and will provide coverage of up to 400 square feet. For larger areas it is usually better to treat each individual work station as if it were an independent room. Sensors should not be mounted lower than 8 feet nor higher than 12 feet above the floor irrespective of the room size or configuration. The amount of coverage is a function of ceiling height.

If the sensor is mounted in a ceiling that is 8′-8″ A.F.F. the radius of coverage will be 6.5 feet for a work plane height of 30″. If the detector is located in a lower or higher ceiling height the radius of coverage will be smaller or larger, respectively. The control unit is composed of a light power switch, a low voltage electronic unregulated power supply and a multi-tap transformer completely enclosed and installed within a NEMA metal enclosure. One control unit is required for each Infracon system.

The sensors should not be installed in close proximity to air registers, or in a direct path of air movement, nor be placed close to a heat pump or other heating or cooling device. Position sensors within 6 feet of work stations but at least one foot away from all fluorescent lighting fixtures and HVAC ducts, and at least 2 feet away from incandescent lighting fixtures. It should be operated in an environment having a temperature range of from 32°F to 120°F where the rate of temperature change will not exceed 20°F per hour. It further requires a humidity range of from 20% to 90% and will work properly in any building that is located at any altitude of from 0 to 6000 feet above sea level.

The complete system will usually pay for itself within one to three years of installation. The payback period is dependent upon the type and size of the facility, occupancy patterns and local utility rates. During warm weather an Infracon system will not only reduce lighting energy consumption but its companion air conditioning energy as well.

FLEXIBLE MODULAR WIRING SYSTEMS FOR INTERIOR LIGHTING CIRCUITRY

As an alternative to hard wiring (conductors in conduit), a flexible, coordinated and integrated modular plug-in electric wiring system that meets all fixture wiring and switching requirements should be considered. Although the material cost of the flexible wiring systems may be higher, the cost of labor for installation is lower, resulting in a substantially lower installed cost. (See Figure 6–28.) The system is more flexible than hard wiring. It allows changes to be made readily without the services of an electrician, while allowing for complete reuse of all system components. This last fact alone reduces the life-cycle cost of tenant changes, space reallocation and relocation. It represents good energy conservation and good material conservation since existing wiring and conduit need not be scrapped and replaced for every tenant change in layout. (See Figures 6–29, 6–30, 6–31, 6–32 and 6–33.)

Life-cycle savings depend upon how often lighting fixture layouts must be changed for renovation and space reallocation. However, a flexible modular wiring system can cut the cost of lighting fixture layout changes by approximately 50%.

The system components include a flexible modular system of wiring for lighting and lighting controls which enables all terminal devices to be connected by flexible cables and specially designed plugs. The plugs are color coded by function, and designed to interface with only the corresponding receptacles of the same compatible voltage system, wiring and switching function.

When planning a new building or modernizing an existing one, consideration should be given to a flexible or modular wiring system, especially for overhead lighting.

Modular wiring permits non-uniform illumination, provides the quantity and quality of the light where, when and as needed and saves energy expenses. Because the luminaires can be disconnected, moved and easily reconnected, this system permits in-house maintenance, design and modification to accommodate changing office situations. Most intriguing of all, because modular wiring systems can be completely and conveniently removed from a building, they are not considered an inherently permanent part of the building but represent tangible personal property. The lighting fixtures that utilize this wiring method are also considered tangible personal property and as such are eligible for an investment tax credit and may be depreciated by one of several accelerated methods.

There are several manufacturers of flexible modular wiring systems that are highly flexible in the options available to the contractor and

	FIXTURE SPACING (CENTER TO CENTER)	AREA PER LIGHTING FIXTURE	HARD WIRING COST	COST PER SQ. FT.	FLEXIBLE WIRING	COST PER SQ. FT.	COST DIFFERENCE PER SQ. FT.
SINGLE LEVEL SWITCHING	4	16 FT^2	$27.73	$1.73	$28.85	$1.80	-$0.07
	6	36 FT^2	$33.25	$0.92	$30.58	$0.85	+$0.07
	8	64 FT^2	$43.09	$0.67	$32.42	$0.51	+$0.16
	10	100 FT^2	$48.61	$0.49	$34.00	$0.34	+$0.15
	12	144 FT^2	$54.13	$0.38	$36.00	$0.25	+$0.13
TWO LEVEL HI-LOW SWITCHING	4	16 FT^2	$29.48	$1.84	$30.69	$1.92	-$0.08
	6	36 FT^2	$35.62	$0.99	$32.58	$0.91	+$0.08
	8	64 FT^2	$46.56	$0.73	$34.89	$0.55	+$0.18
	10	100 FT^2	$52.73	$0.53	$36.78	$0.37	+$0.16
	12	144 FT^2	$58.88	$0.41	$38.67	$0.27	+$0.14

WIRING	MATERIAL	LABOR **
FLEXIBLE MODULAR WIRING	75%	25%
HARD WIRING	25%	75%

** JOURNEYMAN ELECTRICIAN LABOR $30/HR

NOTE:

COST PER FLUORESCENT FIXTURE INCLUDES ALL MATERIALS AND LABOR REQUIRED FOR BRANCH CIRCUIT CONNECTION. COST DOES NOT INCLUDE LIGHTING FIXTURE, FIXTURE INSTALLATION, BRANCH CIRCUIT BREAKER OR PANELBOARD HOMERUN.

Figure 6–28. Comparative cost analysis study based on Figure 6–30's layout.

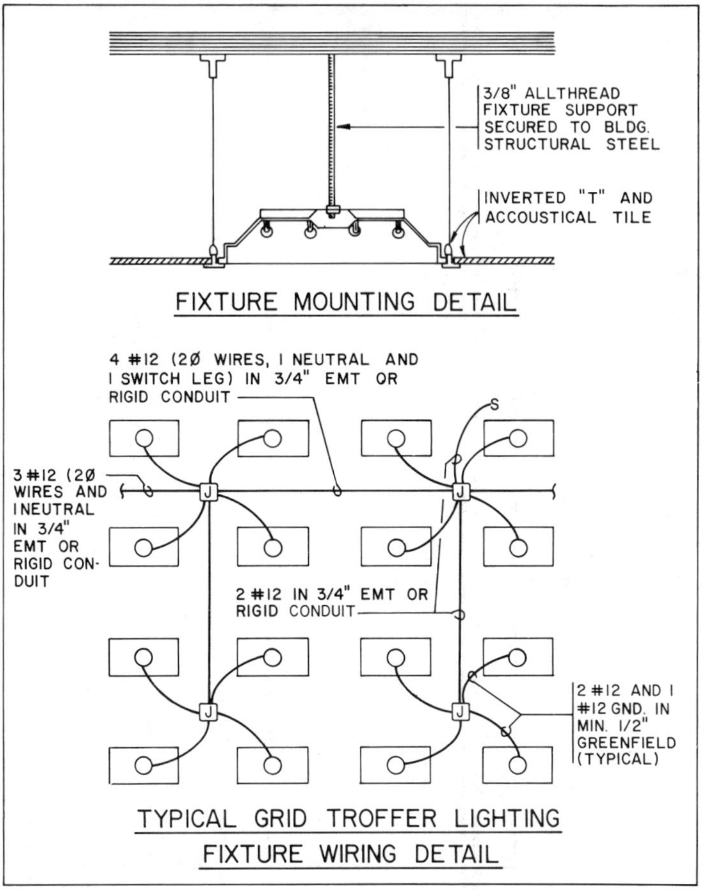

Figure 6–29. Fixture mounting and wiring detail.

owner and thus more expensive than systems with fewer alternative options and less flexibility. Manufacturers that offer highly flexible and sophisticated systems of which some offer lower cost, more speculative systems are as follows: Dual Lite Wiring Products Division, Relocatable Wiring Systems, Incorporated (RELOCK), Electro/Connect Division and Walker-Parkersburg, Walker Flex A.P.S. Other manufacturers of more rigid and less flexible systems are Miroflector and Wiremold. These manufacturers utilize several centralized rigid distribution spline secondary electric service wireways. Into these distribution ducts are plugged flexible wiring whips that are then connected to OEM-installed fluorescent fixture receptacles.

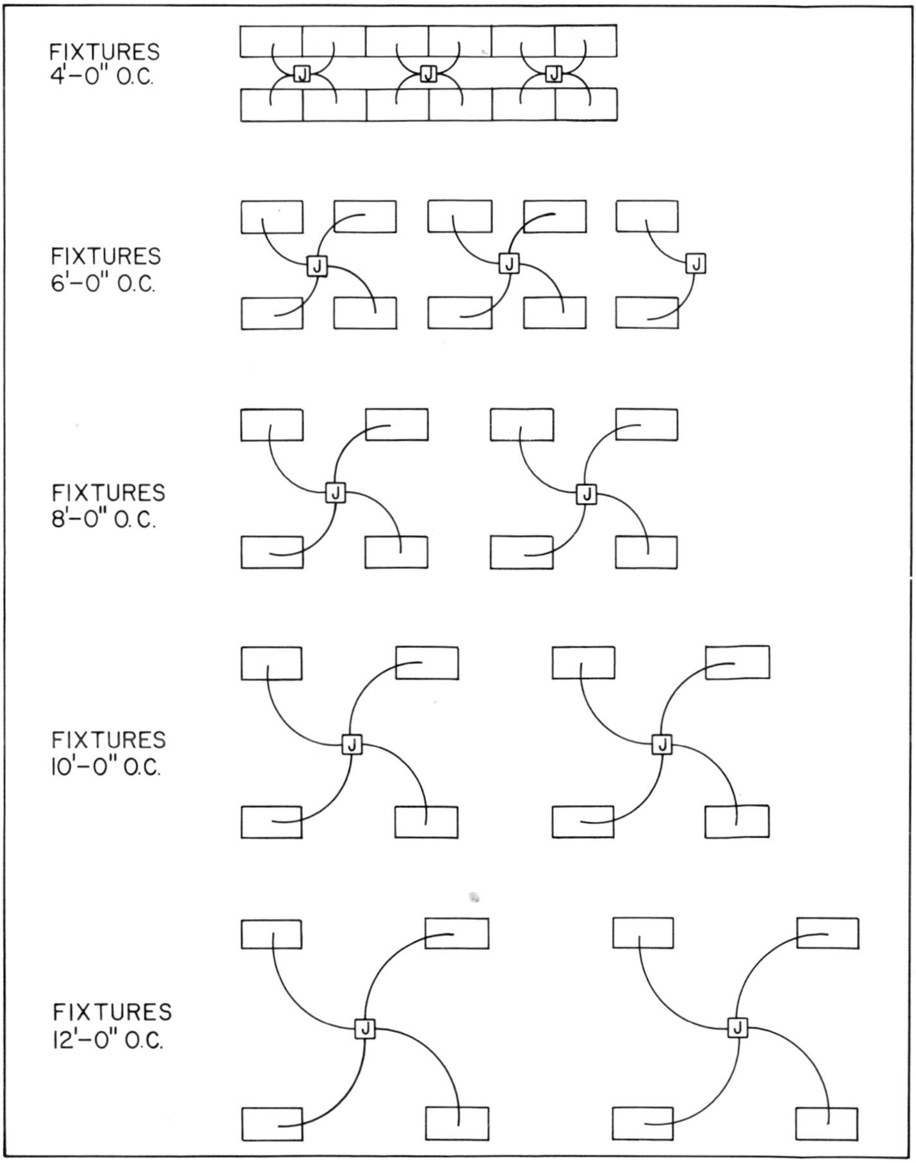

Figure 6–30. Various configurations and spacing layouts for flexible modular wiring system cost comparisons.

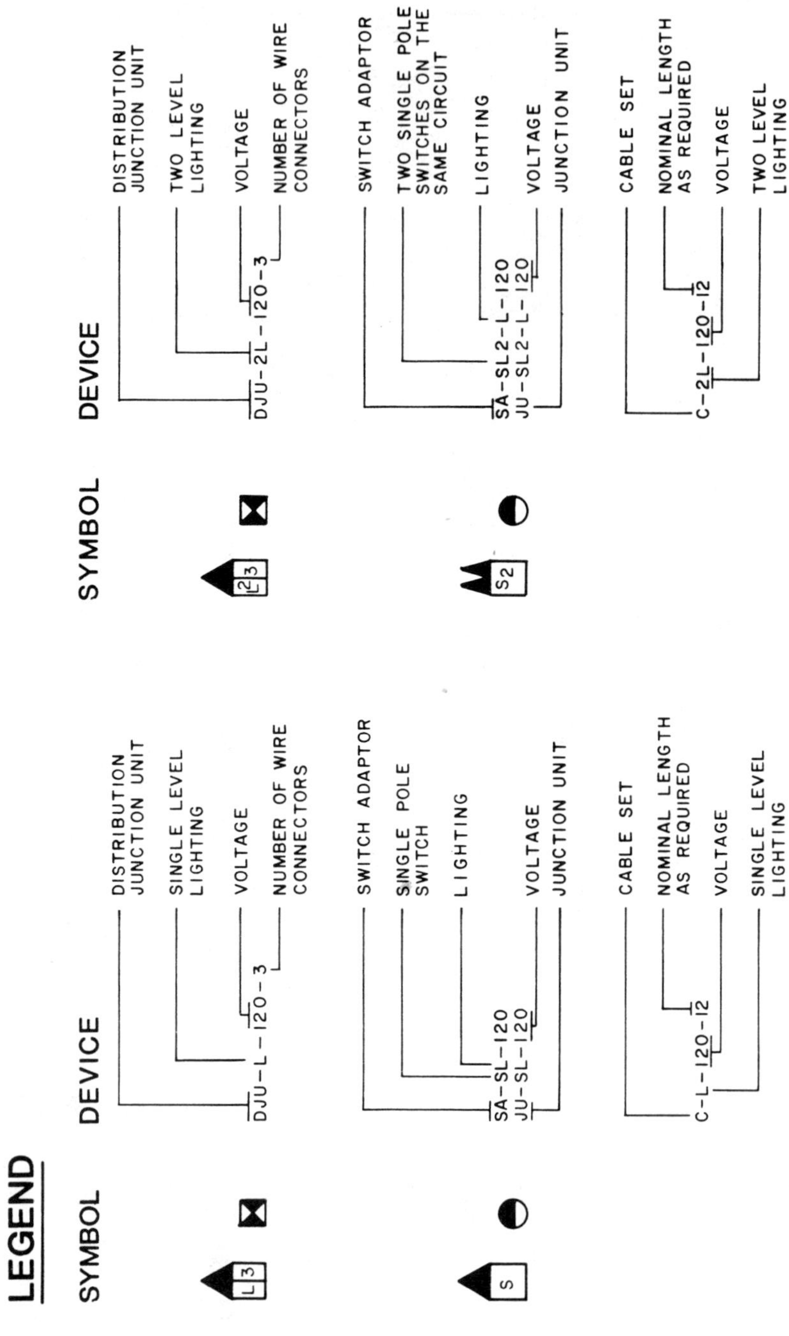

Figure 6–31. Typical flexible modular wiring system symbols.

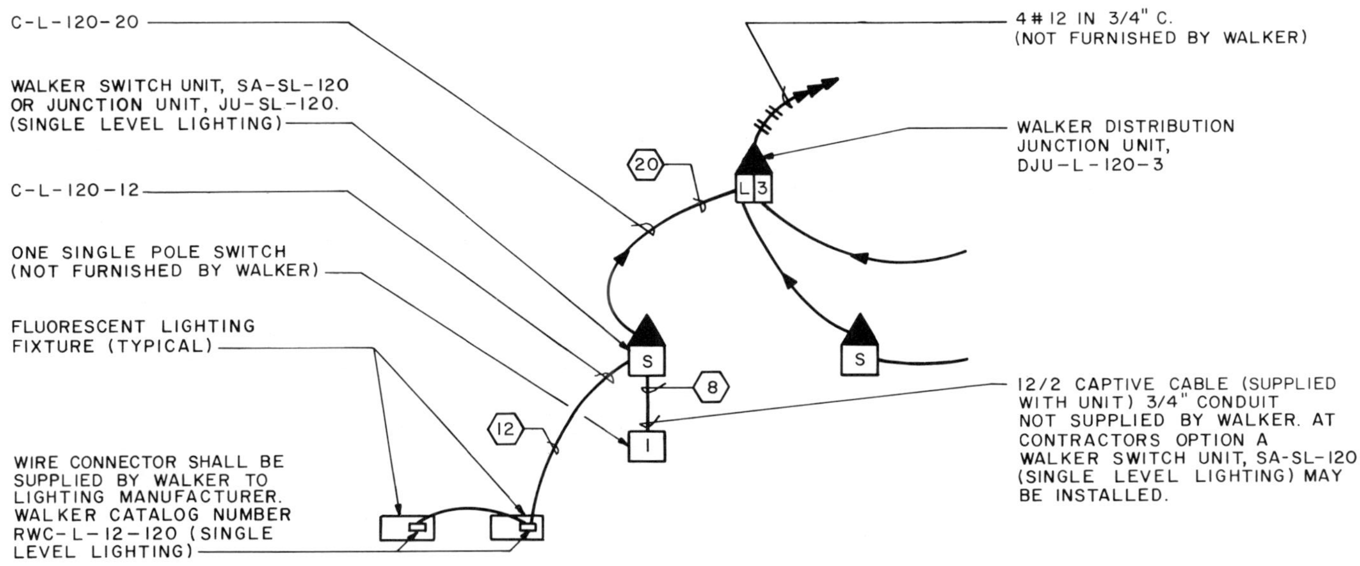

SINGLE LEVEL LIGHTING

SCALE: NO SCALE

SCHEME 1

Figure 6–32. Single level lighting: Scheme 1.

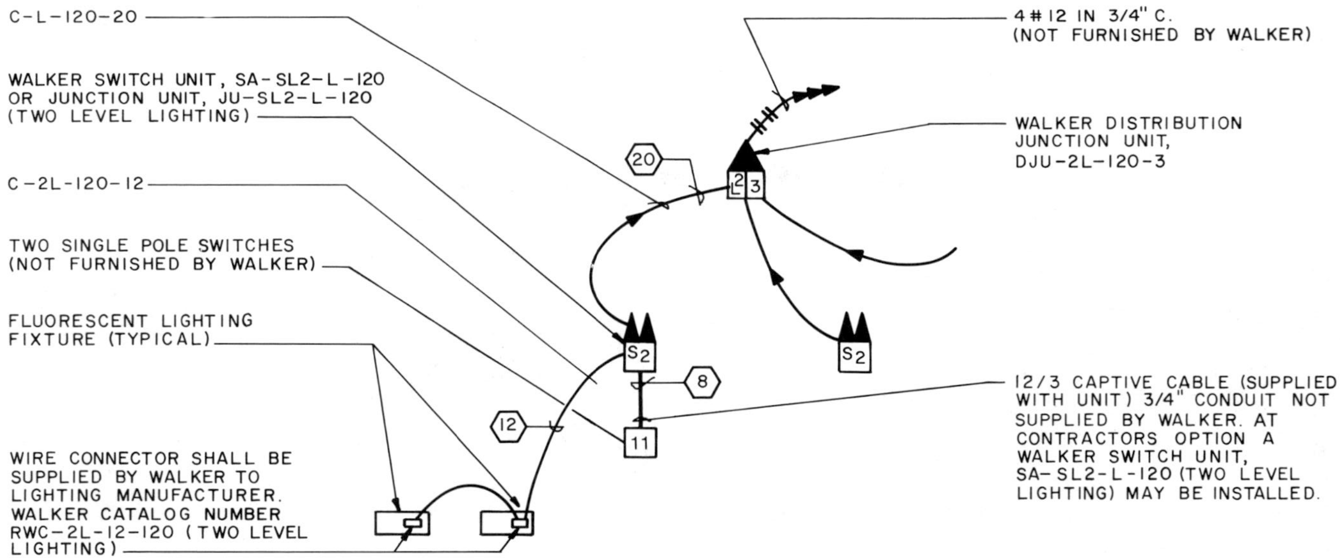

TWO LEVEL LIGHTING

SCALE: NO SCALE

SCHEME 2

Figure 6–33. Two level lighting: Scheme 2.

CONCLUSION

There are already available many new OEM energy saving devices and technologies on the market today. In addition, there is a vast proliferation of new UL listed and labeled technological marvels for new designs and installations as well as for retrofitting into existing equipment that can increase lighting quantity and quality while reducing energy costs. Some are more cost effective than others and have a greater validity to the claim of making a worthwhile contribution to our technology and our society as a whole. To a world going on increasingly shorter rations, the lighting engineer is bringing us tomorrow's technology today, to preserve today's way of life tomorrow. Lamps, luminaires and ballast engineers and technicians, along with control specialists, are helping us to build tomorrow's world today. Truly a new, fascinating future awaits us on the frontiers of lighting technology.

7 ENERGY CONSERVATIVE INTERIOR LIGHTING

INTRODUCTION

Prior to the last decade, it was possible to design interior lighting within an energy conscious framework, but the tools and techniques available to the engineer were fewer and less sophisticated. Although dynamic air handling fluorescent fixtures have been available for over 15 years, and the first high intensity discharge lamp, the mercury vapor lamp, has been available for nearly 50 years, their limitations in the past reduced their usefulness for interior lighting. MV lamps for interior lighting were seldom used until recent years for two reasons: phosphor coated lamps with color corrected and enhanced light output had not been developed, and available fixture designs were limited primarily to roadway and parking lot area luminaires and to industrial fixtures.

Neither was the heat-of-light principle well developed since there was little engineering data available concerning energy to the occupied space, energy to the dropped ceiling cavity space, cubic feet per minute supply and return air capabilities, or even the effect on lamp efficacy (lumens output per watt of power consumed). Various approaches to fixture design were considered, including supply and return air through louvered fixtures and even water-cooled fixtures. The disadvantages of some of these approaches outweighed their benefits. Air supply through the center of louvered fixtures cooled the lamps to the point that light output was actually reduced. Water-cooled fixture systems were expensive to install and maintain, and they were prone to leakage; their lamp compartment temperature was not reduced to the point of substantial benefit to the lamps and ballasts.

These two examples point out the inherent problems to which lighting design, like all technical disciplines, is prone. Lighting technology has now passed beyond the shakedown period necessary to eliminate many of these problems and permit an overall inter-disciplinary integration of various lighting systems' techniques. This chapter will consider some of the basic design concepts which contribute to the realization of effective, efficient lighting at low life-cycle cost.

Fixtures and their component parts may be specified in a number of ways to offer various operating characteristics, different useful lives and hence, different total costs over the life of an installation. In general, departures from "standard" practice can slightly increase the first cost of the installation, while significantly lowering costs over a number of years, due to reductions in maintenance costs and energy use. Specific recommendations for lamp and fixture specifications follow.

LAMP AND BALLAST SEASONING

A fluorescent lamp is a low-pressure mercury, electric-discharge lamp in which a fluorescent coating (phosphor) transforms some of the ultraviolet energy generated by the electric discharge into visible light. High intensity discharge lamps are a general group of lamps consisting of mercury vapor, metal halide and high-pressure sodium lamps. All fluorescent and HID lamps should be "seasoned" to condition them for stable light output and to increase their useful life. The reasons for seasoning are as follows.

Fluorescent and mercury vapor lamps produce 10% to 12% more lumens than their published initial ratings during the first 100 hours of operation. Metal halide lamps produce 18% to 20% more. Also, since a new installation has not been subject to lamp lumen depreciation or lamp and luminaire dirt depreciation, light output will be initially high for these reasons as well. In a relatively clean environment such as a school, office or computer facility, the maintenance factor (MF =

LLD × LDD) for enclosed luminaires is approximately 70%. Thus, if lamps are not seasoned before occupancy of the space, initial light output may be up to 60% higher than the designed level (112 − 70)% × 100/70 = 60%.

For these reasons, all fluorescent and HID lamps should be seasoned. Upon initially having their arcs struck, the lamps should not be turned off until they have burned continuously for 100 hours. This precaution will usually extend useful fluorescent ballast life from a standard 12 to 15 years to as much as 16 to 20 years. Also, seasoning helps to immediately identify faulty ballasts, which can then be replaced while still under warranty, and at no cost to the owner.

HID lamps should be seasoned in their initial burning position. Ideally, fluorescent lamps should be seasoned in a horizontal position. All lamps should be seasoned at rated voltage and on suitable ballasts. No satisfactory method has yet been developed for faster forced-seasoning.

FIXTURE LENSES

Lens selection involves many criteria which the engineer or designer must consider to develop a final lighting design with high energy efficiency (maximum footcandle levels at the task for power consumed), combined with high *visual comfort probability* (VCP). Both qualities are of the utmost importance and neither by itself insures a good lighting design; it is as possible to have a too-bright and uncomfortable lighting environment as to have a comfortable but inadequately lighted environment.

As a minimum standard of quality, all lighting fixture lenses should be specified as 100% virgin acrylic plastic, with a useful life of 15–20 years. Less expensive polystyrene lenses should not be used as they tend to yellow within 2–3 years due to oxygen cross-linking of the mers, with a subsequent reduction in light transmission caused by the larger and heavier atomic weight of the oxygen molecules which also increases brittleness of the lens. The following analysis considers the qualities of lenses in greater detail.

Batwing distribution lenses such as the KSH-3E, KSH-701 Lensmatic and Daybrite VIP offer the advantage of wide spacing-to-mounting-height ratios. This tends to reduce the number of lighting fixtures required if fewer are sufficient to produce the illumination levels required. In a conventional lighting design, there may be instances where, in order not to exceed the spacing-to-mounting-height ratio (S/MH), a greater number of fixtures may have to be used than is required to maintain a given illumination level. (See Figures 7–1, 7–2 and 7–3.) However, if recommended spacing ratios are exceeded, then poor lighting quality will result, with more eye strain and fatigue as a result. In order to maintain the proper spacing-to-mounting-height ratio, there may be certain locations that have higher illumination levels than required if conventional lenses are used in lieu of a batwing distribution type in order not to exceed the maximum-to-minimum illumination levels by more than 1/6. Batwing lenses do permit wider S/MH and thus, in some cases, fewer fixtures may be required to produce the same even illumination level in a given space without exceeding the fixture's S/MH.

Although conventional linear or spherical batwing lenses are capable of producing high equivalent sphere illumination (ESI), they do have their deficiencies. The ideal situation is to orient either single or continuous rows of lighting fixtures (fixture's long axis) perpendicular to the tasks. (See Figure 7–4(A).) In this ideal arrangement, the observer does not receive reflected glare which normally produces veiling reflections and contributes to disability glare. This technique improves the quality of lighting in 50% of the visual orientations possible, where it approximates the illumination produced from a luminous sphere, thus increasing the lighting effectiveness factor.

This result is obtained so long as the lighting fixtures are perpendicular to the task and the observer is looking parallel to the row of light fixtures. However, task locations being impermanent, there are times when the observer is oriented perpendicular to the rows of fixtures (the task is parallel). For the task oriented in this manner, or 50% of the possible visual orientation, lighting is of poor quality due to direct and indirect glare as a result of fixture brightness in a zone above 45°. (See Figures 7–4(B), 7–5 and 7–6.)

In all areas other than critical task areas where recessed fixtures will be used, the lighting fixtures should come supplied with tinted prismatic lenses (such as a KSH K–19 silver tint) which gives a more comfortable visual environment and gives good lamp obscuration. This type of lens represents a slight cost increase over a more conventional lens that does not offer lamp obscuration nor provide for high visual comfort probability (such as a KSH K–12), but the tinted lens does

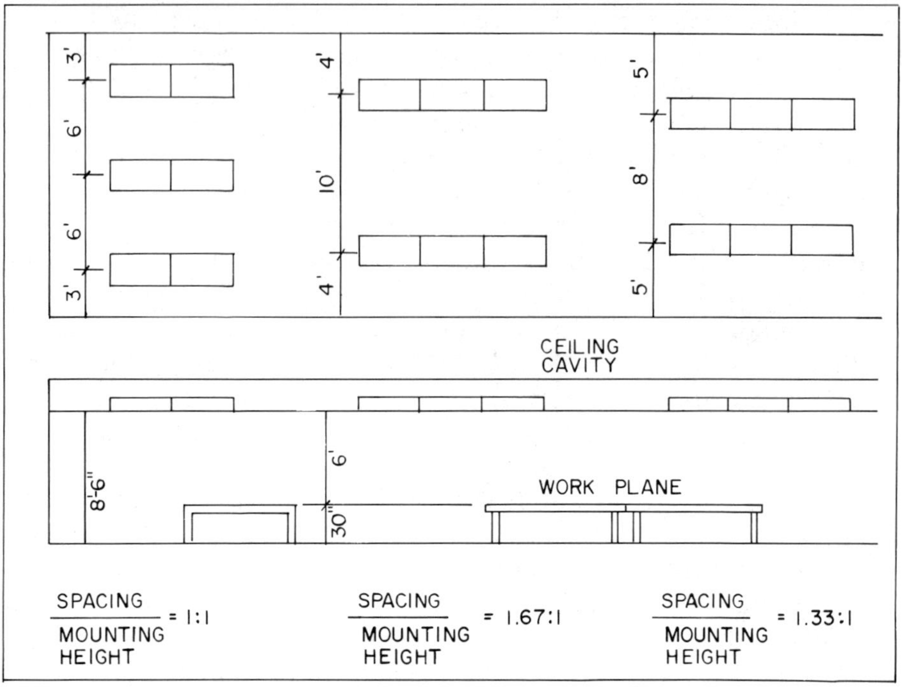

Figure 7-1. Fixture mounting height and spacing ratio.

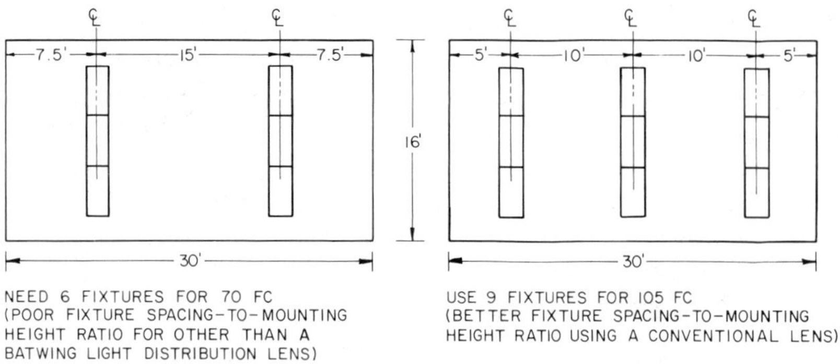

Figure 7-2. S/MH comparison of various lens light distribution patterns. A fixture having a S/MH ratio of 1:35 maximum is installed in a ceiling that is 8′-6″ A.F.F. for a mounting height of 6′-0″.

provide a superior lighting installation in terms of lower direct glare and resulting higher visual efficiency and is well worth the nominal additional cost.

A refractive grid lens with an optional optical interface (such as the Holophane 8224–Y Controlens) should be considered for utilization in all critical task function rooms such as computer rooms and rooms having computer relay terminals (CRTs) or visual display terminals (VDTs). This type of lens increases the VCP to the 80% range, decreases disability glare due to direct glare, and reduces veiling reflections caused by indirect glare. It also decreases high contrast rendition ratios between adjacent areas. Refractive grid lenses are more efficient than tinted lenses, thus requiring fewer fixtures and less energy use for a given illumination level; however, they are much more expensive. This lens in addition to giving high VCPs also produces high ESI, thus increasing the lighting effectiveness factor (Figures 7–5 and 7–7), and is one of the best solutions to the reduction of fatigue in workers by reducing veiling reflections in TV monitors, and desk-mounted CRTs and VDTs. (See Figure 7–8.)

Where fixtures are surface mounted, wraparound lenses are frequently used in order to provide even, high levels of illumination and an aesthetically pleasing cosmetic look. Although these lenses offer the designer fewer options than lay-in lenses for grid troffer-type fixtures, they must still be specified with care. As a minimum, like all fluorescent fixture lenses, they should be specified as being 100% virgin acrylic plastic, and consideration should be given to specifying a lens which meets the *scissors-curve,* direct glare criteria of the Illuminating Engineering Society (IES).

Scissors-curve criteria are used to evaluate average brightness in the zones of fixture light output between 45° to 85° from nadir (point perpendicular to and directly below the fixture), which includes the zones of offending glare. With lenses meeting scissors-curve criteria, higher footcandle levels up to the IES standard, 100 FC can be realized in offices without sacrificing worker comfort due to direct glare or reducing the lighting effectiveness factor. The scissors-curve has been accepted as a standard school lighting criteria in a number of states including California, Georgia, New Jersey, Pennsylvania, Virginia and West Virginia. Lenses meeting these criteria provide a superior lighting installation in terms of lower direct glare and fixture brightness,

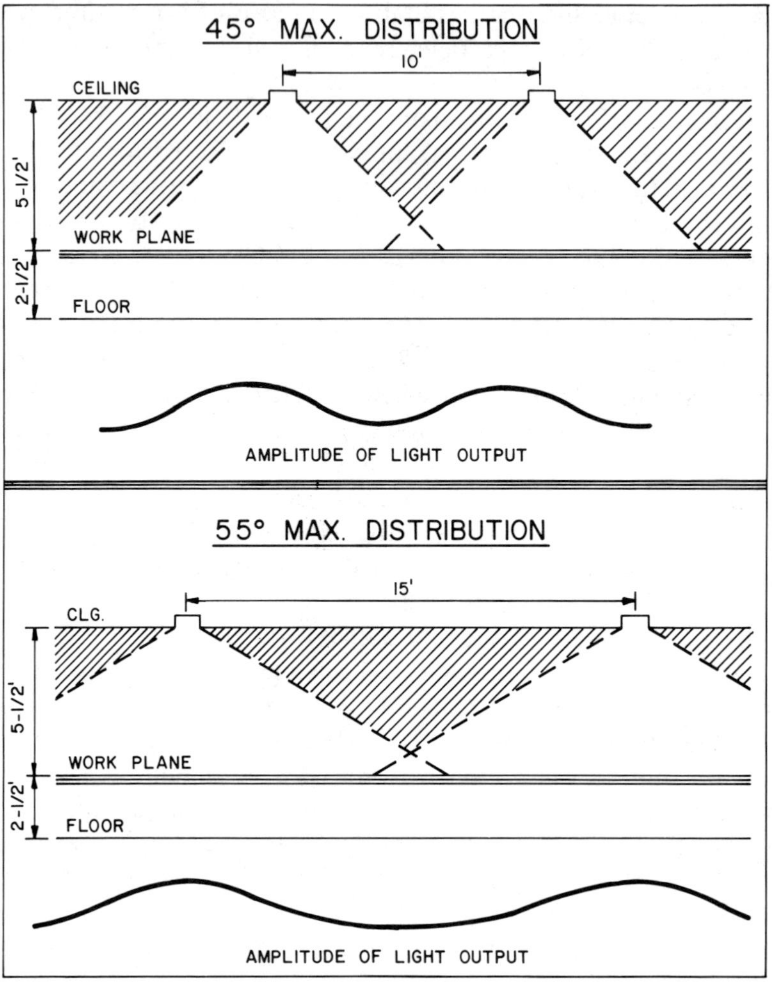

Figure 7–3. Lighting distribution.

with a resulting higher visual comfort probability (up to the 80% range). They decrease disability glare due to direct glare and reduce veiling reflections caused by indirect glare. They also decrease high contrast rendition ratios between adjacent areas, decrease brightness ratios and decrease the illumination ratios.

A typical fluorescent lighting design utilized in an open architec-

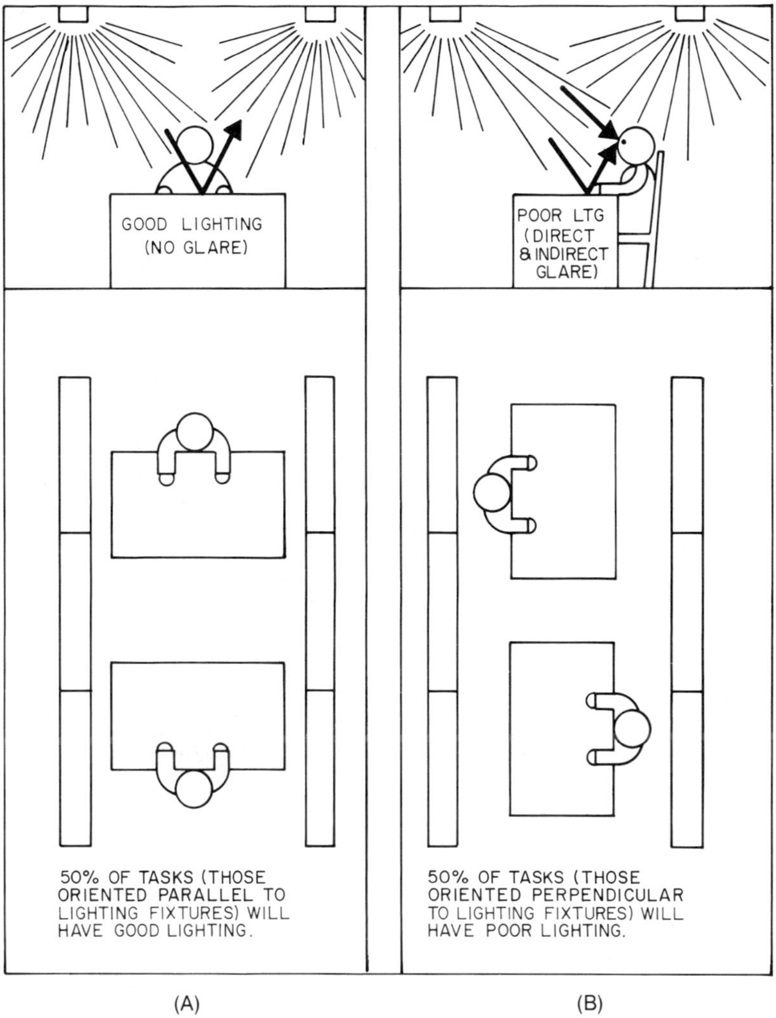

Figure 7–4. Batwing distribution of light.

tural space plan with bank style partitions and employing one 2′ × 4′ 4-lamp, 40-watt lamp fixture per 64 square feet (fixtures on 8-foot centers) will achieve 70 to 75 footcandles (depending on surface reflectances) at 2.875 watts per square foot, or 24 to 26 footcandles per watt per square foot. With four-lamp, direct/indirect lighting fixtures and

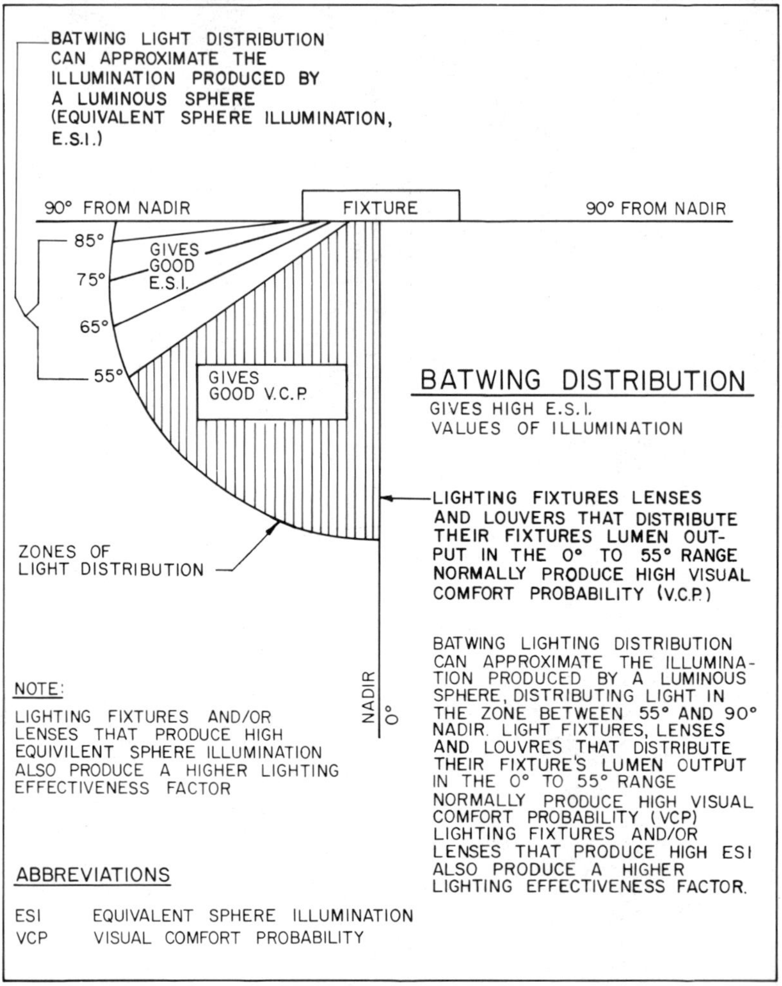

Figure 7–5. Lighting distribution and its relationship to ESI and VCP.

ideal conditions of mounting height and ceiling reflectance, the area per fixture would be reduced to approximately 49 square feet (fixtures on 7-foot centers). Depending on wall and floor reflectances in addition to those of the ceiling, 70 to 75 footcandles would be achieved at 3.75 watts/ft^2 for between 18.64 to 19.97 fc/watt/ft^2.

Lighting fixture types and their patterns of light distribution can

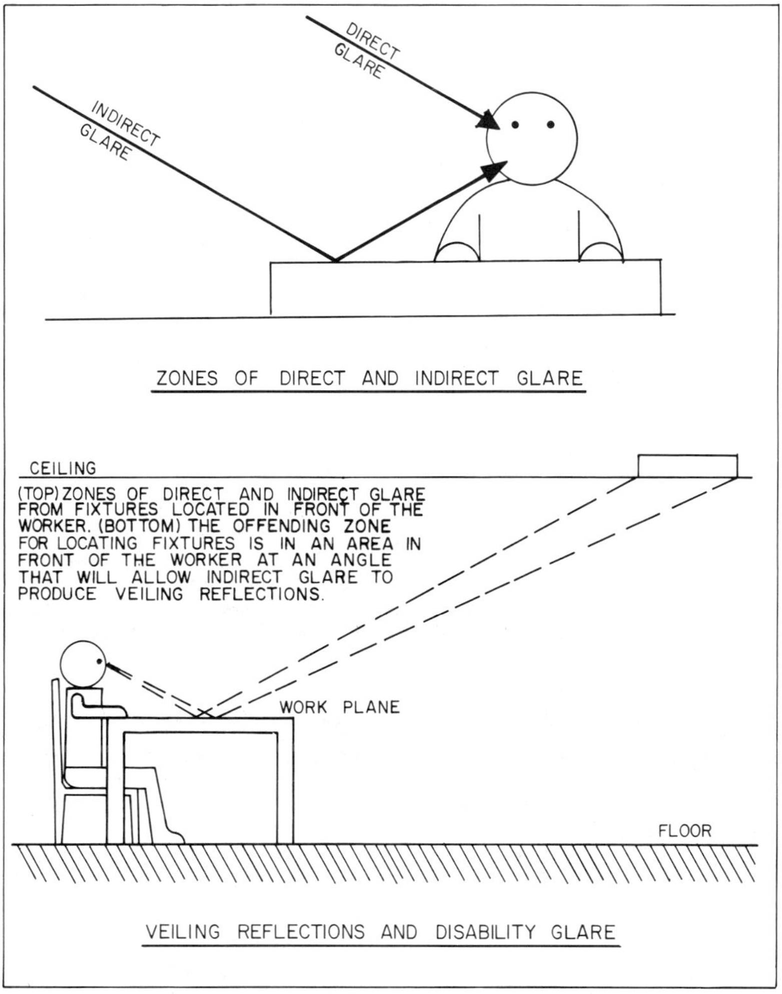

Figure 7-6. Glare analysis.

and do have a large impact on energy consumption levels, as well as a strong impact on illumination levels. As can be seen in Figure 7-9, overhead ambient lighting fixtures producing direct light distribution produce the highest conventional footcandles of illumination for the lowest levels of energy consumed. Their ESI in general will be lower than the several other light distribution approaches shown. However,

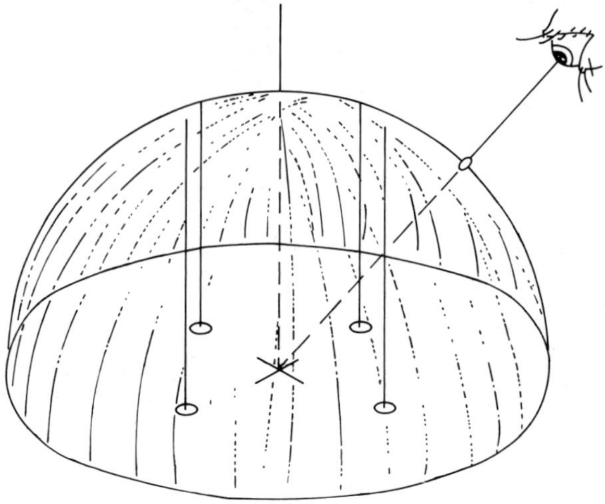

Figure 7-7. Equivalent sphere illumination. Equivalent sphere illumination (ESI) is the illumination produced by a reference lighting situation that will provide the same visibility of a task as when illuminated by an actual lighting system. The reference lighting situation is chosen to be a luminous sphere system, providing uniform lighting from all directions such as at the center of an illuminated sphere, as visualized above.

the proper selection of lenses or louvers can affect this and in general the result is that as the ESI becomes higher so does the energy consumption. A direct-indirect overhead ambient lighting fixture, as a rule, will produce reasonably high levels of conventional illumination coupled with fairly low ESI and higher energy consumption. Indirect overhead ambient lighting units normally consume larger energy amounts with lower levels of conventional fc; however, their ESI is much higher. Wall-mounted valance and ceiling-mounted cove lighting generally consumes the highest levels of energy while producing the lowest levels of conventional illumination but rendering good ESI values.

PROTECTION AGAINST ULTRAVIOLET RADIATION FROM HID SOURCES

The Department of Health, Eduction and Welfare issued a Notice of Alert in November 1977 concerning possible hazards from HID mercury vapor and metal halide lamps. The notice was directed to all

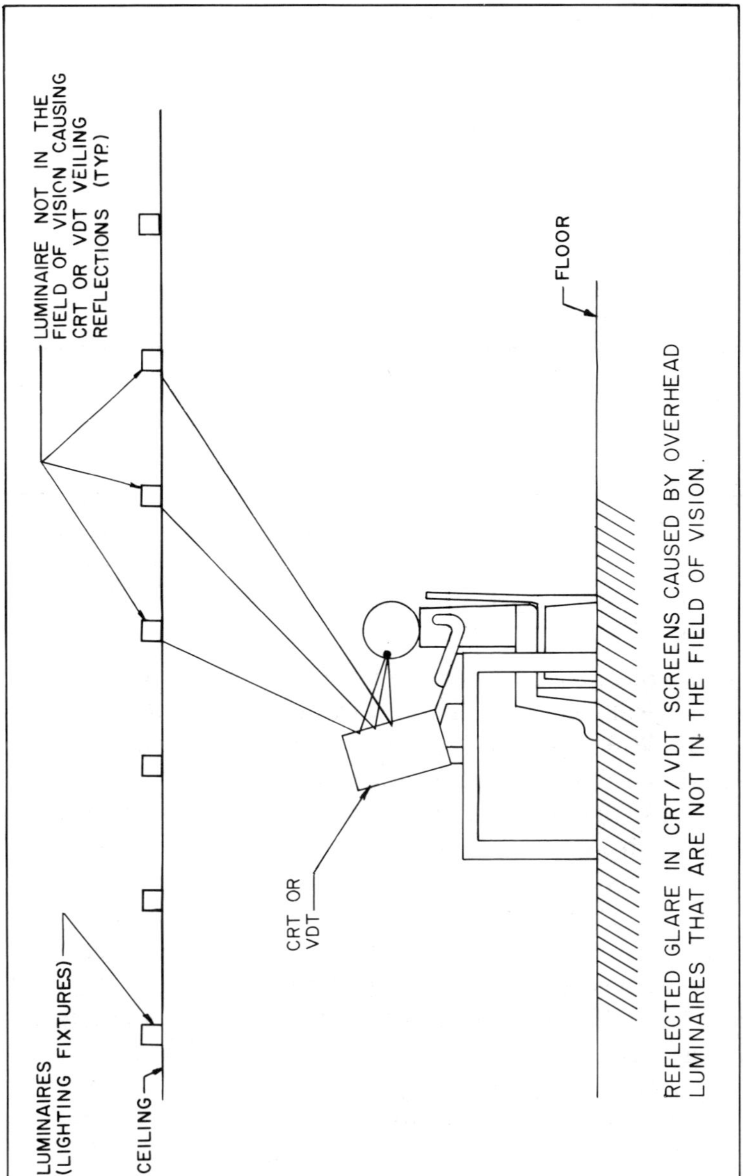

Figure 7–8. CRT/VDT screen viewing reflections.

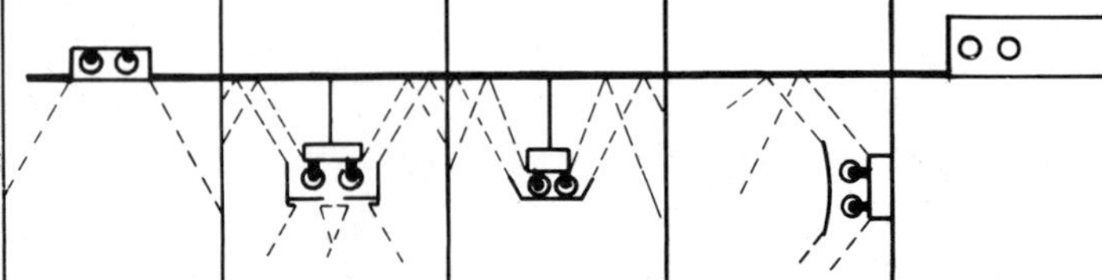

LIGHT DISTRIBUTION AND FIXTURE TYPE	DIRECT	DIRECT-INDIRECT	INDIRECT	VALANCE	COVE
CONVENTIONAL FOOTCANDLES	HIGHEST	HIGH	LOW	LOW	LOW
EQUIVALENT SPHERE ILLUMINATION	LOW	LOW	HIGH	HIGHEST	HIGHEST
ENERGY USE	LOW	HIGH	HIGHEST	HIGHEST	HIGHEST

Figure 7-9. Light distribution.

"Designers of Lighting Installations" and originated with the Bureau of Radiological Health—Food and Drug Administration.

The HEW alert was not a request for a mandatory change, but served as a warning of possible hazards from HID lamps under certain conditions. If the outer bulb envelope of a mercury vapor or metal halide lamp drops away or is broken, people in the area can be exposed to potentially dangerous ultraviolet radiation. HPS and high CRI HPS lamps, as well as low pressure sodium lamps, do not emit UV radiation, as they do not have any mercury vapor amalgam in their arc plasmas. Therefore, the HEW alert does not apply to them. MV and MH HID lamp arcs and arc tubes emit UV radiation, but the outer glass envelope absorbs it and converts it to heat. In the case of phosphor-coated lamps, some of the UV energy is converted into visible light. Prolonged exposure to UV radiation (½ hour) can cause severe skin burn and intense or repeated exposure (10 to 20 hours) may lead to skin cancer. Prolonged staring at an exposed arc tube (15 minutes) due to a broken glass envelope can cause eye irritation and sunburn of the retina (conjunctivitis). This condition can cause temporary blindness lasting up to several days (usually three). In no case has medical history, to date, recorded permanent eye damage.

The Bureau of Radiological Health (BRH) has been aware of the problems involving injuries from broken vapor lamps. BRH published in the Federal Register (Vol. 94, No. 175, September 7, 1979) specifications to protect against accidental exposure to harmful ultraviolet rays, and the following supportive material is excerpted:

a. "Radiation Safety Performance Standard (§ 1040) for high intensity mercury vapor discharge lamps including the metal halide type." The standard is effective March 7, 1980 requiring self-extinguishing of lamp within 15 minutes following complete breakage. After September 7, 1981, self-extinguishing in 15 minutes must follow for even partial breakage, i.e., 3 cm^2 hole or greater. Other standards and requirements are listed in the Standard and it should be consulted for complete details.
b. BRH's letter also confirms the superiority of the built-in safety filament fuse versus the "mechanical" safety switch. All Duro-Test Safe-T-Vapor mercury lamps employ built-in safety filament fuses rather than a mechanical switch.

There are several design approaches to this potential danger. One is to utilize a totally enclosed, glass-lensed fixture; another is to modify open-bottom, downlight fixtures by inserting an interstitial glass lens just under the lamp; a third would utilize a fail-safe extinguishing trigger or filament device. Methods one and two simply convert potentially harmful UV radiation into harmless heat. Method three automatically turns the lamp arc off when the outer envelope is removed.

The automatically extinguishing lamps mentioned in the FDA memorandum are presently available from all major manufacturers, in 100-, 175-, 250-, 400- and 1000-watt mercury vapor. General Electrics 100- and 1000-watt mercury vapor lamps are available on special order. General Electric currently manufactures self-extinguishing 325- and 400-watt multivapor metal halide lamps, and offers it in the 400- and 1000-watt high output line of metal halide lamps. GTE/Sylvania offers this safety feature in the 400- and 1000-watt super metal halide lamps. In addition, Westinghouse manufactures self-extinguishing metal halide lamps in the 400-, 1000- and 1500-watt sizes.

In the Duro Test design, current passes through a heated filament.

When exposed to air after envelope breakage, the filament burns up just like an incandescent filament. Westinghouse, G.E. and Sylvania up until near the end of 1981 used a spring-actuated switch which is compressed shut by the outer envelope. When the envelope is removed, the trigger mechanism opens, extinguishing the lamp's arc. Now Duro-Test, Westinghouse, General Electric and GTE/Sylvania use a heated filament fuse to comply with the latest BRH requirements. The BRH has ruled that any self-extinguishing HID lamps manufactured after September 7, 1981 must extinguish properly when a hole as small as 3 square centimeters is present in the outer jacket. Although several schemes could be designed to meet the new BRH ruling, the filament type device is the only one presently in use. Sylvania self-extinguishing lamps are called Safeline Lamps.

Undoubtedly, the least costly and the best means of protecting against potential UV danger is to utilize an interstitial glass lens to intercept the radiation and convert it to harmless heat. This approach is a one-time cost item good for the life of the lighting fixture. Automatic extinguishing lamps on the other hand, cost substantially more than conventional non-safety lamps. This extra expense will be incurred at every relamping.

The potential problem that has been incurred by the mechanical switch design is that it is not completely fail-safe, according to Robert G. Britain, Director, Division of Compliance, Bureau of Radiological Health, "... since it appears that a mechanical type mechanism which relies upon the entire outer envelope being broken to extinguish a damaged mercury vapor lamp would not have been useful as a safety feature in preventing these injuries...." (BRH cites two incidents where *81* people were injured by only partial lamp breakage.) Therefore, if a fail-safe lamp is to be used, the alternative to a mechanical switch must be the use of a filament fuse type. The only disadvantage to that approach is that a power loss of another 12 watts per each 400-watt HID, MV or MH lamp will be incurred by the tungsten coil. Presumably the same energy wastage of 12 watts due to the tungsten coil will be incurred in 40, 75, 100, 175, 250, 750 and 1000-watt MV lamps, as well as in 175, 250, 400, 1000 and 1500 watt MH lamps. In an era of energy shortages and high energy costs, this would appear to be another bureaucratic boondoggle for Big Brother to protect us against our own wishes and waste additional energy. The small additional safety factor does not seem to outweigh nor justify the additional energy consumption. It is merely a waste of those accrued energy savings that were afforded by the development of the newer, higher efficacy lamps.

CONCLUSION

This chapter has considered in some detail the properties of lamps, ballasts and fixtures lenses, and the avoidance of possible hazards from broken HID lamp glass envelopes. The analyses have stressed maximum performance at lowest costs by minimizing energy consumption and maintenance, emphasizing that it is careful and informed planning on the part of the lighting designer which results in a lighting installation of maximum utility at the lowest life-cycle cost and lowest energy consumption due to good energy utilization.

8 DESIGN OF OUTSIDE LIGHTING SYSTEMS

INTRODUCTION

The design of outside lighting systems for areas such as parking lots, playgrounds and roadways requires a broad knowledge of lighting sources (lamps) and luminaires, as well as a complete understanding of the operating characteristics of various lamp ballasts and luminaire combinations under varying conditions of wind and temperature, lamp lumen depreciation and lamp dirt depreciation. (See Figures 8–1 and 8–2.) Outside lighting design, be it for security, safety or both, requires certain basic and intrinsic insights into these characteristics. By way of introduction, consider first that basic modern light source, the fluorescent lamp, made commercially available by General Electric in 1938.

Fluorescent lamp light output is a function of mercury vapor pressure which is usually controlled by the coolest spot on the bulb wall; rated lumen output values are the lumens obtained at 77°F ambient air. Light output is reduced at both high and low temperatures. (See Figure 8–3.) For maximum light output (optimum mercury pressure), the coolest point on the bulb should be about 100°F. As the bulb wall temperature is increased, mercury pressure increases and light output decreases. At lower bulb wall temperatures, mercury condenses, the mercury pressure drops and light output decreases.

Figure 8–1. Washington frosted foods and cold storage warehouse and office space with street and facade lighting. MV and MH light sources were utilized.

Figure 8–2. Diamond Elementary School parking lot and driveway lighting. Pole luminaires utilized 400-watt SMH lamps while entrance downlights utilized 175-watt MV lamp sources.

Figure 8–3. The Washington Metropolitan Area Transit Authority's (WMATA) L'Enfant Plaza bus terminal. From above the light coming from the tinted skylight dome appears to be incandescent lighting.

Figure 8–4. From underneath it becomes apparent that the source of illumination for the WMATA L'Enfant Plaza station is fluorescent.

Thus, the performance of a fluorescent lamp is affected by the type of lighting fixture it is installed in, its location (indoors or outdoors) and the type of ballast used. (See Figure 8–4.) All of these factors affect bulb temperature, and all are in turn affected by the application, design requirements and economic factors involved.

Selection of lamps of the highest efficacy (lumens delivered per watt consumed) reduces the number of fixtures required, with subsequent savings in labor for installation and maintenance. Add to these savings the reduction in energy costs which can be realized through careful planning and design, and a total lighting system of low life-cycle costs and maximum utility will result.

Fixtures and their component parts may be specified in a number of ways to offer various operating characteristics, different useful lives and hence different total costs over the life of an installation. In general, departures from standard practice may slightly increase the first cost of the installation, while significantly lowering life-cycle costs over a number of years, due to reductions in maintenance costs and energy use. For a parking garage facility, for example, the following specific lamp and fixture specifications would be recommended.

LAMPS

Straight 40-watt, rapid-start flourescent lamps in the warm white series have 3200 lumens (80 lumens per watt efficacy). These lamps have 20,000 hours' life rated at 3 hours per start, 28,860 hours' life rated on 10 hours per start, and 37,700 hours' life on continuous burning, at a cost of $2.85 each. The ballast and lamp together consume 54 watts in a single lamp installation. A two-lamp, 40-watt fluorescent ballast consumes 92 watts of power, or a 17.4% energy savings compared to two single-lamp fixtures. Thus, wherever possible 2-lamp ballast fixtures should be used. The use of VHO lamps is not recommended for the following reasons:

- VHO lamps are only 85% as efficient as conventional 40-watt, rapid-start lamps.
- An 8-foot, 2-lamp VHO industrial fixture costs approximately 150% as much as the two 4-foot, 2-lamp, 40-watt lamp fixtures it replaces.

- Two 215-watt VHO lamps cost aproximately 275% as much as the 4-lamp, 40-watt, rapid-start lamps they replace.
- VHO lamps have a useful life of one-half that of 40-watt, rapid-start lamps.
- VHO ballasts are more costly and have a shorter life than 40-watt, rapid-start ballasts.

Employment of VHO lamps is thus more costly on both a first cost and a life-cycle cost basis (increasing energy costs plus greater expense for maintenance materials and labor) than a design using conventional 40-watt rapid start lamps.

FIXTURE BODIES

One-piece, injection-molded, high-impact, thermoplastic bodies are an excellent choice for parking structure lighting fixtures. Alternately, bodies with an outer housing of one-piece polyester reinforced, self-extinguishing fiberglass having a continuous neoprene gasket should be considered equal to thermoplastic bodies in quality and performance. Plastic and fiberglass fixture bodies are also impervious to damage from automobile aerials.

Aluminum bodied lighting fixtures normally should not be considered for an outdoor project. Although aluminum forms a protective coating of aluminum oxide that does not further oxidize, this coating causes the paint to peel off in the same way that it does on rusting steel bodied fixtures. If plain polished aluminum lighting fixtures are used, their initial 88% or lower reflectivity rapidly deteriorates to 40% or 45% due to oxide formation, and this reaction decreases the useful light output by 50% or more. Anodized aluminum bodies do not oxidize nor does their light output deteriorate appreciably; however, the first cost of these fixtures is usually too high to consider for a parking garage.

Steel bodied lighting fixtures are more economical than aluminum fixtures from both first cost and life-cycle costs. In addition, the newer five- and seven-stage zinc phosphate coating and hot-bonded, baked-enamel finish, or a porcelain enamel finish over cold-rolled steel, are processes that virtually eliminate rusting under damp conditions. However, porcelain enamel finishes are more prone to mechanical damage. Since all ballasts produce a *leakage current* which can build up a charge on a steel fixture and cause bodily injury if a pedestrian were to touch the fixture while standing on a wet floor, the ballasts and fixture should be properly grounded. (Grounding also increases the life of the ballast.) Plastic bodied fixtures, of course, preclude the possibility of electrical shock.

LENSES

Fixture lenses should be of the wraparound type, with gaskets for moisture and insect protection, and should be specified as 100% virgin acrylic plastic with a high degree of ultraviolet resistivity, having a useful life of 15 to 20 years. Less expensive polystyrene lenses should not be used as they tend to yellow within 2 to 3 years with a subsequent reduction in light transmission and an increase in brittleness due to the larger oxygen molecules crosslinking with individual polymer strands. The specification of fixtures with lenses and gaskets insures maximum performance of the fluorescent lamps by maintaining the lamp ambient operating temperature well within the optimum range while reducing the wind chill effect which could retard lamp starting in cold weather.

According to Westinghouse, "standard lamps enclosed in sealed luminaires produce their highest output between 0°F and 50°F; in open luminaires, they produce highest output above 50°F." Data from Arm-a-Lite, which manufactures protective plastic sleeves for fluorescent lamps, indicates that a shielded F40 lamp produces the same light output between 32°F and 80°F, while at lower temperatures a shielded lamp will perform better than unprotected lamps. (See Figure 8-5.) G.E. points out that an enclosed, gasketed, F40 standard lamp with 0°F ballasts within a sheltered structure (eliminating the wind chill factor since wind cannot blow over the luminaire) will not be subject to abnormal lumen depreciation; lamp output will remain essentially the same at 0°F as it was at 70°F, since by maintaining lamp bulb temperature, optimum mercury vapor pressure is maintained within the tube. Even in the extremely adverse conditions found in areas such as Chicago, Milwaukee and Green Bay, Wisconsin, Kenall lighting fixture data indicate that with enclosed fixtures, lamp efficiencies are not reduced to unacceptable levels.

Use of enclosed lensed fixtures also eliminates the need to install −20°F high-output lamp ballasts, which are more expensive than 0°F

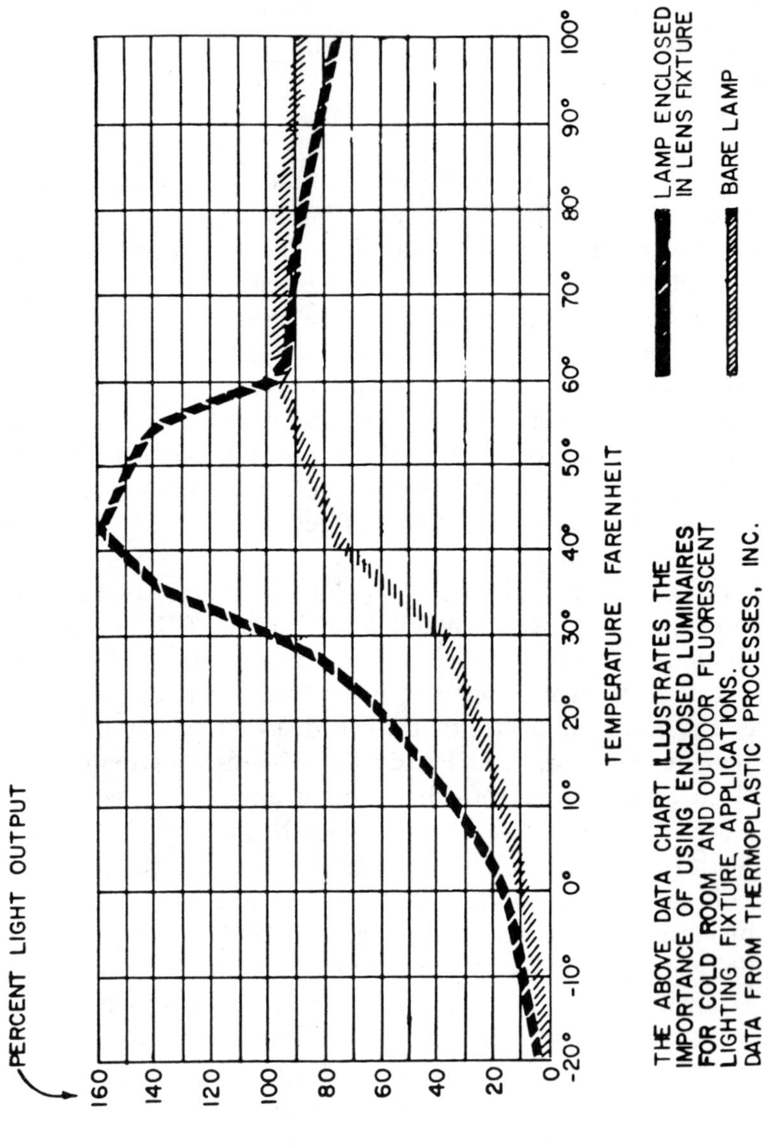
PERCENT LIGHT OUTPUT
160
140
120
100
80
60
40
20
0
-20°
-10°
0°
10°
20°
30°
40°
50°
60°
70°
80°
90°
100°
TEMPERATURE FARENHEIT
LAMP ENCLOSED IN LENS FIXTURE
BARE LAMP
THE ABOVE DATA CHART ILLUSTRATES THE IMPORTANCE OF USING ENCLOSED LUMINAIRES FOR COLD ROOM AND OUTDOOR FLUORESCENT LIGHTING FIXTURE APPLICATIONS.
DATA FROM THERMOPLASTIC PROCESSES, INC.

Figure 8–5.

rapid-start lamp ballasts. Data from Kenall indicate that within enclosed luminaires, 0°F ballasts have proven suitable for the coldest applications.

SECURITY AND ACCENT LIGHTING

Security and accent lighting is frequently designed more from an aesthetic than from a photometric sense, with walkway and planter lighting and tree uplighting seen as adequate for required levels of area illumination. However, if security is an important consideration, dependence upon spill illumination is not sufficient. (See Figure 8-6.) The amount of general illumination realized from tree uplighting is small at best, and decreases during the winter when the deciduous trees lose their foliage. Also, spotty lighting sometimes can reduce visibility due to the high contrast between lighted and dark areas. Generally, for security lighting, a contrast ratio of not more than 6:1 (light:dark) should be maintained.

Thus, spotlighting is generally less effective than diffuse illumination. Washing vertical walls to form a broad luminous surface is particularly effective since it results in silhouetting which highlights not

Figure 8-6. Smithsonian Museum's Hirshorn Sculpture Garden utilizes the concept of silhouette security lighting by washing vertical walls with illumination. Here the bare winter trees are silhouetted by broad floodlight back lighting. As can be seen here, any spill illumination from tree uplighting would be nil during this period of the year.

only sculptures, planting or architectural features, but unauthorized persons as well. (See Figures 8–7, 8–8 and 8–9.) Silhouetting is a nearly ideal security lighting technique. (See Figures 8–10 and 8–11.)

This was the approach studied in the renovation of the Smithsonian Museum's Hirshorn Sculpture Garden. Mercury vapor floodlights with an unusually wide beam pattern, or light-spread, were to be mounted at ground level, and normally would have been aimed up at the north and west walls, which would have tended to accentuate the exposed aggregate with an even wash of grazing light. At the south wall, the fixtures were aimed instead at the soffit formed by the stairs above. Reflected downlight from the soffit, in conjunction with the fixture uplight, provides particularly even illumination at this wall, and increased illumination at ground level. (See Figure 8–12.) The south wall fixtures are to be concealed behind new benches. At the reflecting pool and cul-de-sac walls the fixtures were to be concealed either in light troughs running along the walls or in new bollard-type seats.

The step light fixtures used along stairs and ramps also utilized mercury vapor lamps, and they are concealed in concrete curbs. These lighting fixtures have a wide beam pattern to minimize the number of fixtures required and also to minimize their energy consumption. (See Figure 8–13.)

Figure 8–7. Martin Marietta Corporate Headquarters is utilizing facade lighting by the use of pedestal ground mounted 400-watt super metal halide parabolic reflector floodlight fixtures.

Figure 8–8. The author's son, John Zackrison, enacting the part of the intruder and vandal. Here he is partially illuminated by spill illumination and thus he can be identified.

Figure 8–9. In this scene, John stands out only in silhouette.

Figure 8–10. Pedestal mounted mercury vapor floodlights illuminate the area around Martin Marietta Headquarters as well as the underside of the soffit in order to form a visually distinctive band separating the cafeteria area and lounge below from the corporate executive floor areas above.

Figure 8–11. Pedestal mounted wide beam pattern mercury vapor flood luminaires produce a particularly even wash of illumination.

Figure 8–12. The spill illumination from the walls in conjunction with the reflected overhead soffit lighting increases the illumination at the ground which provides additional fill illumination of the wall giving it a more evenly lit appearance.

Figure 8–13. 100-watt mercury vapor wall recessed curb isle lighting fixtures produces even illumination of the handicapped stair ramp due to its wide beam spread.

Figure 8–14. Wall and/or stair illumination by the use of fluorescent handrail lighting fixtures. Photo courtesy of Controlight.

Figure 8–15. Good, even levels of security and safety lighting of the handicapped stair ramp are obtained at low levels of energy consumption.

Alternate design concepts were also explored. In one, the reflecting pool and cul-de-sac walls might have been lighted with continuous rows of narrow-profile, rail-type fluorescent fixtures (see Figure 8–14) mounted at the tops of the walls, or set out from the wall on stanchions to provide very even illumination over the entire extent of the walls.

The operating costs of providing this additional lighting have been minimized through proper lamp selection. Both the mercury vapor and fluorescent lamps offer long life (24,000 hours or more), reducing the costs of maintenance. Total energy use for stair ramp and security lighting as now designed and installed will not exceed 7.2 kW or 0.15 watts/square foot. (See Figure 8–15.) This is a much lower level of energy consumption than would be possible if incandescent fixtures were used for overall security lighting and provides a better overall quality of illumination.

GENERAL SITE LIGHTING CONSIDERATIONS

This chapter has so far considered two outdoor lighting applications —a parking structure and security/accent lighting—in depth to illustrate the detailed understanding of lighting equipment that is necessary to arrive at the best solution for each lighting problem. We will now broadly consider more general concepts which form the basis of informed outside lighting design.

Lamps

Incandescent lamps at 15 to 18 lumens/watt efficacy and short lives of several thousand hours have existed for just over a century, since Thomas Edison first tested his carbon filament lamp on October 19, 1879. General Electric introduced the first commercially available mercury vapor lamp in 1934. High pressure sodium lamps were first made available commercially in 1961, (although G.E. did some experimental work on them in the late 1940s), and metal halide lamps became available in 1965.

For many years, then, MV was the most efficient lamp in general use for outdoor lighting, with anywhere from 34 to 61 lumens per watt, depending on wattage size. Clear MV lamps were widely used for site and roadway applications, since as a point source they were easily controlled by reflectors, refractors or combinations of the two (as in the standard *cobra head* type of luminaire). Also, since the MV lamp, like all HID lamps, has an outer glass envelope enclosing the arc tube, the arc maintains its temperatures under cold conditions; hence fixtures did not have to be selected to protect the lamp from temperature lumen depreciation. This made MV lamps considerably more useful than the still more efficient fluorescent lamps for outdoor applications. (Also, fluorescent lamps are a linear light source which makes their light output more difficult to distribute and control effectively.) The clear MV lamp did present the drawback that its characteristic spectral power distribution was bluish-green in color, but later developments in phosphors led to lamps such as the deluxe white, which had a more complete SPD and rendered colors more faithfully.

Further development of HID sources by the inclusion of metal additives has since improved the SPD, efficacies and rated lives of the metal halide family to the point that the lighting design engineer has many capabilities and many options open to him. In particular, outdoor lighting does *not* have to be cut back to realize energy and dollar savings. Numerous advances in the field of lighting system and component design and manufacture now make it possible to have outdoor lighting systems that are more effective, more efficient and far less costly to operate and maintain than ever before. It is now possible to reduce lighting energy costs while improving lighting effectiveness and increasing lighting illumination levels.

Table 8–1 provides information on selected typical lamps. Note, however, that efficiency should be considered in terms beyond those of measured lumens per watt. A full analysis of lamp performance will indicate, for example, that lumens depreciate and wattages actually change (and may increase for some high pressure sodium lamps) during the life of the lamp. In addition, the cost and useful life of the lamp must be considered in order to develop a meaningful cost comparison.

Lamps selected for site lighting should generally be clear rather than phosphor-coated in order to approximate a point source for the most efficient distribution of light. HID lamps with a phosphor coating are not point sources, since the lamp envelope rather than the arc tube becomes the source, and light from large area sources cannot be controlled by fixture reflectors as accurately as that from point sources. For this reason, mercury vapor lamps which are phosphor coated to increase light output and improve the color of light normally should not be used for most highway and site lighting applications.

Table 8-1. Lamp data, © 1977, Gardco Manufacturing Inc. Author's amendations are marked by *.

LAMP TYPE		WATTS (Nominal)	LAMP ORDERING NUMBER		INITIAL LUMENS	LUMENS PER WATT	RATED LIFE
METAL HALIDE	CLEAR HORIZONTAL	1500		M1500/BU-HOR	150,000	100.	1,500
		1000	G.E.	MV1000/U	107,800	107.8	10,000
		1000		M1000/BU-HOR	95,000	95.	10,000
		400	SYL.	MS400/HOR	40,000	100.	•20,000
		400		M400/BU-HOR	•34,000	•85.	•20,000
		250		M250/BU-HOR	19,500	78.	•10,000
		175	SYL.	MS175/HOR	15,000	85.7	7,500
	CLEAR VERTICAL	1000	SYL.	MS1000/BU	125,000	125.	10,000
		1000	G.E.	MV1000/BU/1	115,000	115.	10,000
		1000	West.	MH1000/BU	100,000	100.	10,000
		400		M400/BU	34,000	85.	15,000
HIGH PRESSURE SODIUM		1000		LU1000	140,000	140.	24,000
		400		LU400	50,000	125.	24,000
		310	G.E.	LU310	37,000	119.4	24,000
		250	G.E.	LU250/BU/S	30,000	120.	24,000
		250		LU250	25,500	102.	24,000
		150		LU150/55	16,000	106.7	24,000
		150		LU150/100	16,000	106.7	24,000
		100		LU100	9,500	95.	•24,000
		70		LU70	5,800	82.9	•24,000
DELUXE WHITE MERCURY		1000		H1000DX36-15	60,000	60.	24,000
		400		H400DX33-1	21,750	54.4	24,000
		250		H250DX37-5	12,300	49.2	24,000
		175		H175DX39-22	8,180	46.7	24,000
		100		H100DX38-4	4,200	42.0	24,000
•MET HAL. CLR. VERT		400	SYL.	MS400	40,000	100.	15,000

UPDATING FOOTCANDLE VALUES

$$\frac{\text{New Lamp Lumen Output}}{\text{Old Lamp Lumen Output}} = \frac{\text{Updating}}{\text{Multiplier}} \qquad \text{Example: 400 W Metal Halide } \frac{40000}{32000} = 1.25 \text{ Factor}$$

Old Footcandle Value x Updating Factor = New Footcandle Value

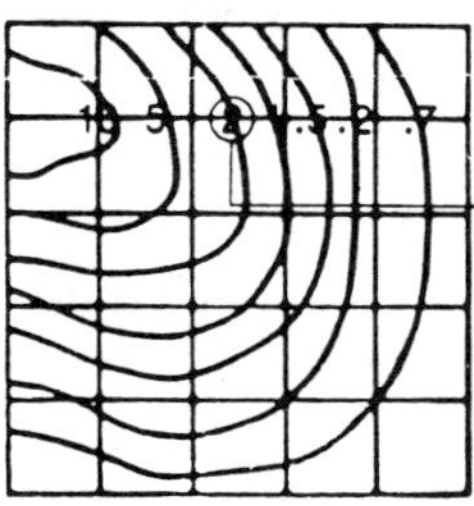

Example:

2FC x 1.25 = 2.5 Initial Footcandles (Based on M400 rated at 40000 lumens).

EXISTING FOOTCANDLES (Based on M400 Rated at 32000 Lumens).

Mercury vapor lamps also have lower efficacies than other HID lamp sources.

Super metal halide lamps have very high efficacies of 100 and 125 lumens/watt for 400- and 1000-watt units, and very good color rendering qualities; and the clear SMH lamp is ideally suitable as a point source in all types of luminaires. High pressure sodium lamps are also small sources, although they produce a primarily dichromatic light, sometimes called a *golden white* color. As such, they tend to distort some colors. This is a relatively minor drawback when the lamp is used for general outdoor illumination, but when used for parking lot lighting, their use tends to make colors other than orange or yellow appear as shades of gray, except for reds which appear as a rusty brown color. Car identification by color at nighttime can thus become difficult if not impossible. HPS has very high efficacies of from 82 to 140 lumens/watt depending on lamp wattage (with the notable exception of the smaller wattage sizes such as the 35-watt HPS lamp at 64 lumens per watt), so high in fact, that when it is retrofitted into older cobra-head luminaires, the luminaire which was never intended to control such a high output lamp becomes a source of glare.

Low pressure sodium lamps are the most efficient lamps available today, delivering from 137 to 183 lumens per watt depending upon lamp wattage. Although they were first developed in this country in the late 1920s, they have been used much more extensively in Europe for the past 50 years. LPS produces a monochromatic yellow light which tends to render all colors except yellow as shades of gray. Further, it is a linear light source, hence it is difficult to control and distribute the lumen output efficiently. Although LPS is finding applications in this country, the lamps at present are all imported, which could lead to problems of availability or quality control.

When using different lamp types together on the same project, consideration should be given to matching their SPD for aesthetic purposes. One should consider the following lamp groupings:

FLUORESCENT	METAL HALIDE	MERCURY VAPOR
Warm white	Phosphor coated	Warm deluxe white (see chart below)
Cool white	Clear	Brite white deluxe (see chart below)

EQUIVALENT MERCURY VAPOR LAMPS

MANUFACTURER	LAMP DESIGNATIONS		
General Electric	Delux white	Warm deluxe White	Color Improved
Sylvania	Brite white deluxe	Warm tone	Color Improved
Westinghouse	Deluxe white	Style tone	N.A.

Pole Heights and Fixture Design

Varying pole heights is another technique which can reduce energy costs. When taller poles are used, it enables the use of higher wattage lamps which usually are more efficient than their smaller wattage counterparts (see below). As shown in Table 1, for example, a 1000-watt high pressure sodium lamp is about 12% more efficient than the 400-watt version. At the same time, using taller poles with higher wattage lamps in luminaires that provide good directional control generally means that fewer lamps and poles are needed. Note, too, that taller poles do *not* necessarily mean increased maintenance headaches, since they are available with luminaires that can be lowered by an electric drill or drive motor attached to a winch device at the bottom of the pole. Cranes, skyhooks and bucket trucks (cherry pickers) are not needed at all.

Consideration must be given to the type of luminaires involved. Those selected should be able to control objectionable light spillage and glare that could affect adjacent areas. Luminaires that provide sharp cut-off type distribution not only reduce spill light onto adjoining property, but minimize direct glare into the eyes of oncoming motorists or pedestrians. (See Figure 8–16.) High mast luminaires also serve to reduce glare since at long distances the relative brightness

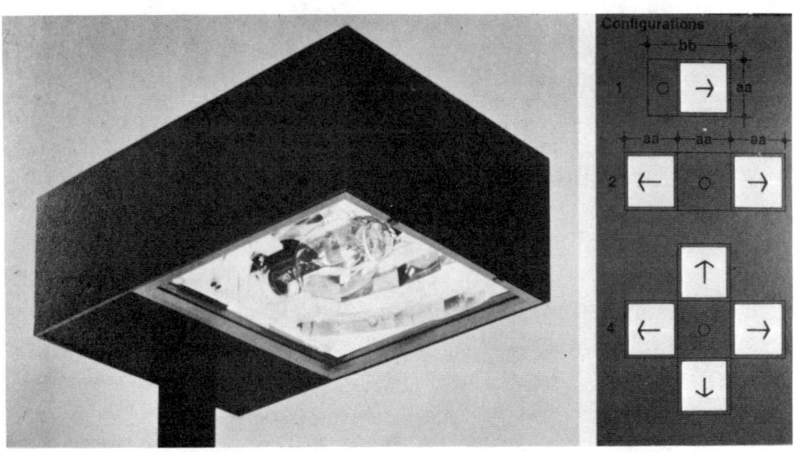

Figure 8–16. High angle sharp cutoff high pole height luminaire produces good, even illumination at distances up to 6 times the pole height. There are several configurations for this type of pole-mounted luminaire, single, double or quadruple mounting.

(contrast) is low, and therefore not disturbing. Close to the mast, when the luminaire is almost directly overhead, direct glare is cut off by the roof of the car.

Globe-type luminaires are probably the least efficient and most glare-producing of all, since light is distributed in all directions 360° in two planes, rather than down upon the ground. (See Figure 8–17.) However, inclusion of a refractor around the lamp both effectively directs illumination upon the area intended and serves to diminish the source of glare rather than illuminate the foliage and clouds above. (See Figure 8–18.)

POLE HEIGHT STUDY FOR PARKING LOT LIGHTING

The parking area was to be illuminated to a minimum of two horizontal fc's is to permit surveillance by closed circuit television. Twelve-foot poles were originally considered for reasons of aesthetics. Using such short poles, the lamp wattage had to be low in

Figure 8–17. This clear globe luminaire has an exposed lamp having no refractor to redistribute the illumination downward to ground level where it belongs. This is an example of a not to functional, primarily aesthetics only, street fixture.

Figure 8–18. A clear globe luminaire with a refractor more effectively redistributes the light onto streets, driveways and parking lots. The top of the refractor is enclosed in order to redirect the light downward more energy efficiently.

order to reduce the total source brightness. So each 12-foot, pole-mounted fixture would use a 250-watt clear MH lamp. Due to the low wattage and efficiency, 80 of these luminaires would have been required to illuminate the parking lot with a minimum of two horizontal fc's required for the closed circuit television. The approximate cost of the pole and luminaire was $450, plus $600 for wiring, raceways, pole bases, circuitry and labor. Each pole fixture would consume 290 watts of electric energy. The 250-watt clear MH lamp has a life of 10,000 hours and costs $49.75.

On the other hand, a 25-foot, pole-mounted lighting fixture can utilize two luminaires, each with a single 400-watt clear SMH lamp. This unit not only uses two lamps per pole; but, by utilizing higher-wattage, higher-efficacy lamps, only 33 of these pole fixtures were required in order to produce the required two horizontal fc's. The approximate cost of the pole and luminaire was $950, plus $600 for wiring, raceways, pole bases, circuitry and labor. Each pole fixture consumed 940 watts of electric energy. The 400-watt clear SMH lamp has a life of 20,000 hours and costs $47.00.

Using the same twin 400-watt SMH luminaire on 30-foot poles, only 18 fixtures were required in order to produce the required horizontal fc due to the better utilization of the light spread. The approximate cost of the pole and luminaire was $975, plus $600 for wiring, raceways, pole bases, circuitry and labor. Each pole fixture consumed 940 watts of electric energy. The 400-watt clear SMH lamp has a life of 20,000 hours and a cost of $47.00.

Economic Analysis

Using 20,000 hours as a base comparison time of usage since it is the lowest common denominator, total lamp replacement labor and material cost for the 12-foot pole fixtures is $9,560. The lamp replacement labor and material cost for the 25-foot pole fixtures over 20,000 hours is only $3,762, a savings of $5,798 over the 12-foot pole fixtures. For the 30-foot pole fixtures, the cost is $2,052; or $7,508 less costly to maintain than the 12-foot fixtures and $1,710 less costly than the 25-foot fixtures.

The first cost of installation is $84,000 for the 12-foot poles, $51,150 for the 25-foot poles and $28,350 for the 30-foot poles. The 25-foot pole lighting fixtures thus cost $32,850 less than the 12-foot pole fixtures from a first cost standpoint. The 30-foot pole lighting fixtures cost $55,650 less than 12-foot and $22,800 less than the 25-foot pole fixtures.

The cost of energy to operate the 12-foot poles over a 20,000 hour time span would be $20,416; the cost of energy over the same time span to operate the 25-foot pole fixture would be $27,297; the cost of energy over the same time span to operate the 30-foot pole fixture would be $14,889. Thus, the cost of energy to operate the 25-foot fixture is the highest by $6,881 over a 20,000 hour span. The cost to operate the 30-foot fixture would be $5,527 less than the 12-foot and $12,408 less than the 25-foot pole fixtures.

The difference in cost to maintain the 12-foot fixture over 20,000 hours (5.48 years) is $1,413 less than the total maintenance cost for the 25-foot fixture. For the same 20,000 hours (5.48 years) the 30-foot fixture is $12,855 less costly to operate than the 12-foot, and $14,268 less costly to operate than the 25-foot fixture.

The difference in initial costs of a 25-foot pole and a 12-foot pole would be amortized by a 12-foot pole in 39 years, if its replacement life was not limited to 15 or 20 years. If a system does not amortize in 7 to 8 years, it is not usually considered economically viable. The 30-foot

LAMP	LAMP EFFICACIES EFFICACY (WITHOUT BALLAST LOSS)	LIFE EXPECTANCY
250 watt MH lamp clear	82 lumens per watt	10,000 hours
250 watt MH lamp phosphorcoated	82 lumens per watt	10,000 hours
400 watt SMH lamp clear	100 lumens per watt	20,000 hours
400 watt SMH lamp phosphorcoated	100 lumens per watt	20,000 hours

EFFICACY IN LUMENS PER WATT AND LIFE OF LAMP WERE CURRENT VALUES AT THE TIME OF THE STUDY IN 1978.

ECONOMIC ANALYSIS AND COST COMPARISON	
12-foot Pole Lighting Fixtures:	
80 fixtures @ 290 watts each	
Initial first cost	$ 84,000.00
Connected load	23.2 kW
Energy consumption per month 10 hrs/day, 30 days/mo	6,960 kWh
Cost of energy per month * $.044/kWh	306.24
Material cost of lamps per 20,000-hr. period	7,960.00
Labor cost per 20,000 hour period @ $10.00 per lamp change	1,600.00
25-foot Pole Lighting Fixtures:	
33 fixtures @ 940 watts each	
Initial first cost	$ 51,150.00
Connected load	31.02 kW
Energy consumption per month 10 hrs/day, 30 days/mo	9,306 kWh
Cost of energy per month @ $.044/kWh	409.46
Material cost of lamps per 20,000-hour period	3,102.00
Labor cost per 20,000-hour period @ $30 per lamp change (twin luminaire)	990.00
30-foot Pole Lighting Fixtures:	
18 Fixtures @ 940 watts each	
Initial first cost	$ 28,350.00
Connected load	16.92 kW
Energy consumption per month 10 hrs/day, 30 days/mo	5,076 kWh
Cost of energy per month @ $.044/kWh	$ 223.34
Material cost of lamps per 20,000-hour period	1,692.00
Labor cost per 20,000-hour period @ $30 per lamp change (twin luminaire)	540.00

* Cost of energy at the time the study was performed in 1978 was $.044/kWh.

pole fixture from both a first cost and a life-cycle cost is more cost-effective (least costly) to own and operate than either a 12-foot or a 25-foot pole fixture.

On the basis of this analysis, the use of twin luminaires using 400-watt clear SMH lamp luminaires mounted on 30-foot poles clearly offers decisive advantages from the standpoints of both first cost and life-cycle cost.

OUTDOOR LIGHTING CIRCUITRY

Two factors are usually taken for granted in wiring outdoor lighting fixtures without being considered in proper depth. One is the utilization voltage; and the second, the methodology of wiring as it relates to voltage drop. Site lighting is frequently served at 120 volts and wired in discrete circuits (3-phase wires sharing a neutral with up to 1,920 watts of lighting load per circuit). However, this requires that the feeder be sized for the voltage drop incurred by a two-way run—the phase conductor out and the neutral conductor back. Because of the long wiring runs common in outdoor lighting, this method results in voltage drop problems of greater magnitude than those common in interior lighting.

Consideration should be given instead to using 3-phase, 277 V circuits to serve outdoor lighting. With HID lamps, all fixtures can be placed on the 480Y/277 V service. Figure 8–19 shows the advantages: there are two wires to the last lighting fixture in the group (phase and neutral); three wires to the next fixture (2 phase wires and a neutral); and four wires to the remaining lighting fixtures (3 phase wires and a neutral). This permits the first fixture on the 3-phase circuit to be on phase A, the second on phase B, the third on phase C, the fourth fixture on phase A, etc. Thus, the only imbalanced neutral current is that contributed by one fixture when there is either one extra fixture on a given phase or one extra fixture on each of two phases. Any condition of even numbers of lighting fixtures on all phases provides zero vector neutral current aside from any third harmonic distortion or contribution.

Even in large site lighting and street lighting installations such as industrial parks, medical and college campuses, lighting should be served at 277 V if the overall system can be fed in sectors from different buildings. However, in the case of a large centrally-fed lighting system, voltage drop at 277 V may be a serious problem requiring ex-

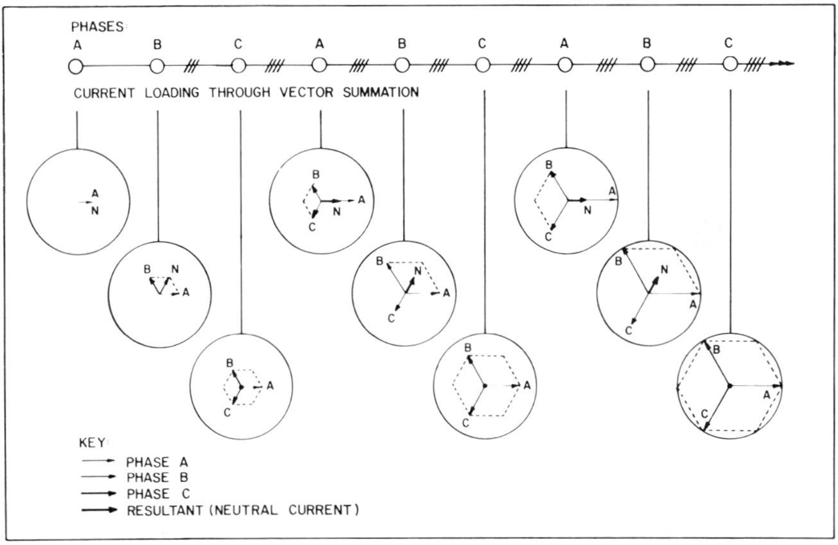

Figure 8–19. Site lighting.

cessive oversizing of direct burial conductors and oversized raceways for those conductors where they pass under roadbeds or paved areas.

At this point it becomes practical to consider the use of a higher distribution voltages for site lighting. Invariably, a large campus or industrial facility will employ the use of central chillers fed by a 4,160Y/2,400-volt, 3-phase, 4-wire distribution system. Thus, site lighting can be wired for a series-multiple 2,400 V distribution.

If incandescent or quartz lamps (which have low efficiencies and short lives, thus higher first and life-cycle costs, as well as higher energy expenditures) are selected, each lighting fixture should be provided with a fused disconnecting pothead in the pole base. Otherwise, since a 2,400 V lighting system is normally wired in series multiple, an entire circuit of lights would go out when a single lamp fails. However, with a fused disconnecting pothead, a transient surge voltage is created when a lamp burns out, melting the fused link in order to bypass the burned-out lamp. This preserves the circuit so that the remaining lamps continue to operate.

If HID luminaires are used, the 2,400 V primary, 135 V secondary ballasts will act as transformers; thus, if a single HID lamp fails it does not open-circuit the remaining fixtures. Minimum starting voltage at −20°F is 225 volts either RMS or peak and its operating voltage is 135 volts.

Today, the most commonly used intermediate range distribution voltages are either 13.2 kV or 34.5 kV. The preferred distribution voltage for industrial parks and college campuses is 13.2 kV (or, for some utilities, 12.47 to 13.8 kV) in order to reduce first cost of distribution equipment, such as electrical feeders and primary switchgear and primary transformers. In the installation where a 2,400/4,160 V chiller and a series multiple-street lighting system are used, a step-down transformer would be required. This will increase first and life-cycle costs while causing additional energy waste in the form of transformer core and coil losses. If, for any reason, the step-down transformers serving the chillers are shut down at any time during the season, the 2,400 V service is unavailabel for outdoor lighting.

Further, if the chillers come on line during the period the lighting is on, a high surge of current through the transformer could create a large voltage drop in the transformer windings, causing a temporary lamp lumen depreciation or light flicker from the street lights. The HID luminaire lamp arc plasma may extinguish if this voltage dip is more than 35% during windless warm weather, exceeds 25% when the outside ambient temperature is 20°F, or the outside temperature is 35°F and there is a stiff wind blowing, causing a lower wind chill-factor equivalent temperature.

This condition would leave an installation in the dark anywhere from 6 to 14 minutes, depending on wind and outdoor temperature conditions, until at least some of the lamp arcs are restruck; however, they would not reach full intensity for another three or more minutes. If, however, a lamp restrikes and then extinguishes, another 5- to 10-minute cooling period is required before the arc plasma is cool enough to attempt a restrike.

AUTOMATIC LIGHTING CONTROL

One way of cutting back on the energy consumed for outdoor lighting is through reliable automatic controls. In some cases, all outdoor lighting is operated manually by one switch located at a central panelboard. It is the least expensive first cost type of control to install, but perhaps one of the most expensive to operate. Lighting may be turned on early, wasting energy; or it may be turned on late, creating a security risk. In any event, it lacks the precision that automatic controls can provide. And since precision implies that lighting will be used only when and where it is needed, it also implies minimizing energy use. Automatic controls are of two types: *time clock and photocontrol.*

Time clock controls activate and deactivate lighting at predetermined times. They provide a higher degree of reliability and are available with an *astronomic* feature which automatically compensates for the seasonally changing hours of darkness, precluding continual adjustment of "on" and "off" times. Time clocks can also be provided with 16-hour reserve power which carries the clock automatically through power outages or brownouts (reductions in voltage) and thus prevents loss of accuracy. A day-omitting feature is also available to prevent automatic operation of lights on weekends and holidays.

Photocontrol switches activate lighting not at preset times but at a predetermined low level of light. Thus, on overcast days, when nightfall appears to occur earlier than usual, lighting will be activated earlier than usual. When the amount of light reaching the optical control recedes to a certain level, the lighting system is turned on automatically. And when the amount of daylight increases, the lighting is extinguished automatically. As such, photocontrol switches do not have to be adjusted from season to season, and they also respond to variations that can occur during a particularly overcast day. This system by itself can be energy wasteful by always operating the lighting system during darkness, even at such times in the early morning or during the daytime on overcast days when it may not be needed. This system should be employed only for lights that are used for security purposes and so should be on whenever natural light is below a predetermined level.

The most energy conservative and thus economical control system (from the standpoint of life-cycle costs) is a hybrid, using a time clock in conjunction with a photocontrol. For most applications, full sunset-to-sunrise lighting is not needed; perhaps sunset-to-2 A.M. is all that is required for security purposes, instead. The time clock is set to turn the lighting on sometime before sunset and off at the required time, while the addition of the photocell switch in series with the clock serves to fine tune the precise minute at which the lights come on—earlier than usual on overcast days, for example, as previously mentioned. A marriage of these two controls, then, permits optimum control by incorporating the advantages of each. The result is lighting when, and only when, needed, and avoidance of unnecessary lighting, energy consumption and cost.

And going even one step further, a **combination of the astronomic dial time switch and the photocontrol** can be used, even though they would seem to perform essentially the same function. This system should be used when it is known that lights should be allowed to come on a half-hour or so before sunset on inclement (overcast) days but should not be allowed to come on during the normal daytime even though the natural light may go below a predetermined "on" level. In other words, in this particular situation there is presumed to be no vandalism risk except during normal nighttime darkness hours. Such fine tuning can conserve enough energy to pay for the small additional cost in a very short time.

It is important when marrying time switches and photocontrols to recognize that the time switch motor must be independent of the photocontrol circuit, so that it will remain operative even though the photocontrol circuit is in the "off" mode.

Today, of course, we must look at the cost of materials, energy and energy reduction through the various techniques of life-cycle costing. In essence, life-cycle costing is a technique that considers all costs of owning and operating something over its useful life. This includes the cost of installation labor, materials and equipment, as well as the costs of operating maintenance and energy usage. By analyzing these costs, it will be found in many instances that spending more on a new system will be worthwhile.

There are many different steps that can be taken to reduce expenses without degrading the quality or quantity of illumination. And if reduced illumination results in increased vandalism, which numerous studies suggest is exactly what will occur, then where are the real savings?

CONCLUSION

This chapter has considered in some detail those factors which affect the design of outdoor lighting systems, with a strong emphasis upon value engineering, energy management and energy conservation. The analyses of lamp and fixture types, site lighting circuitry and automatic controls are intended to contribute to the design of lighting installations which offer maximum performance at reduced energy levels of consumption and minimize first and life-cycle costs.

9
LIGHTING METHODS COMBINING ENERGY CONSERVATION WITH AESTHETICS

INTRODUCTION

Today, the design of lighting systems offers the engineer unmatched opportunities to incorporate in the design of building systems both energy conservation and low life-cycle costs for the owner. The tools necessary to realize these dual objectives are primarily new and improved lamps used in conjunction with new methods of controlled lighting. These methods have only recently reached that state of development at which the designer can now be confident of their consistent performance, and of the accuracy and sufficiency of the manufacturers' engineering data on which his design depends.

This chapter will consider past projects which illustrate energy-conservation problem solving within a framework of both new and old technology, but technology in all cases finely integrated. No single technique can be used to solve all lighting problems; nor can any single building normally employ every technique from considerations of cost, aesthetics, preexisting design assumptions or, in the case of rehabilitations, preexisting conditions. Yet, within these basic limitations (and they are common to every trade and engineering discipline), the quantity and quality of lighting was maximized and the consumption of energy minimized through careful choices between and utilization of what *was* available to the design engineer in each situation.

Building designs and lighting systems should be integrated so as to achieve attractive and energy-efficient illumination. Lighting technology has taken quantum leaps in the past decade, but the successful integration of lighting into building systems, from both aesthetic and energy-conservative standpoints, is as important as ever. It is particularly important that, from the perspective of rising energy costs, energy management and energy conservation not be sacrificed to aesthetic considerations.

Never before has the successful marriage of these two concepts been more possible. Imaginative and energy-conservative lighting designs are right at hand—a diversity of highly efficient lamp types and standard off-the-shelf fixture designs unparalleled in the past. The following buildings were chosen to illustrate the degree to which lighting and building design can be welded into attractive and energy-conservative systems through the use of current technology now available to all designers.

NATIONAL SAVINGS & TRUST, WASHINGTON, DC

A building of striking design, winner of the Oliver Owen Kuhn Award for architectural excellence, the NS&T project, located in Washington, DC, was a challenge to the lighting designer. The irregular form and high cathedral ceiling of the main banking area, energy conservation and an overall integration of the lighting into what was already seen as an exceptional building were important considerations. No less important was the owner's concept of lighting as advertising; the building was to look warm, friendly and inviting at night. (See Figures 9–1 and 9–2.) Hence, the interior lighting had to be carefully considered as an element of the building night lighting. (See Figure 9–3.)

Figure 9–4 is an overview of the outside of the building at night and shows how lighting can effectively advertise a bank. The walk-up window and docutel area (24-hour automatic teller machine) are to the left and are illuminated by three low-glare downlights having 175-watt Bright White Deluxe mercury vapor lamps. The National Savings & Trust sign above the stairwell to the right is illuminated by a 100-watt Brite White Deluxe mercury vapor lamp, medium-flood fixture, pedestal-mounted on top of the stairwell tower.

Figure 9–1. Drive-up window and remote teller station are illuminated by 15-175W bright white deluxe mercury vapor downlights. The access driveway to the left is illuminated by the use of 175W MV wall mounted wall packets.

Figure 9–2. The evenness of illumination is attested to by the overlapping light that is scalloping high up on the wall and by the same relative illumination levels throughout on the ground of the drive-up area.

Figure 9–3. Tellers' area with mezzanine office space with overhead.

Figure 9–4. Night lighting with advertising objectives. The NS&T building is attractively displayed at night.

As a result of several considerations, it was decided early in the design to utilize high intensity discharge sources for the general area, the office and for decorative lighting. Fluorescent fixtures are used only in the vault, the janitor's closet and the mechanical equipment room.

The lamps used are metal halide and mercury vapor. As both phosphor-coated MH and Brite White Deluxe MV lamps offered excellent color rendition and compatibility, the decision concerning which type to use for which application was based on energy conservation and life-cycle costs.

Thus, general area lighting is provided with nine 400-watt, MH, down-light fixtures mounted in the sloping ceiling over the main banking area as shown in Figures 9–3, 9–5 and 9–6.) At 85 lumens/watt and 15,000 hour life (now 20,000 hours), this MH lamp is among the most efficient of the HID lamps available for general illumination. The

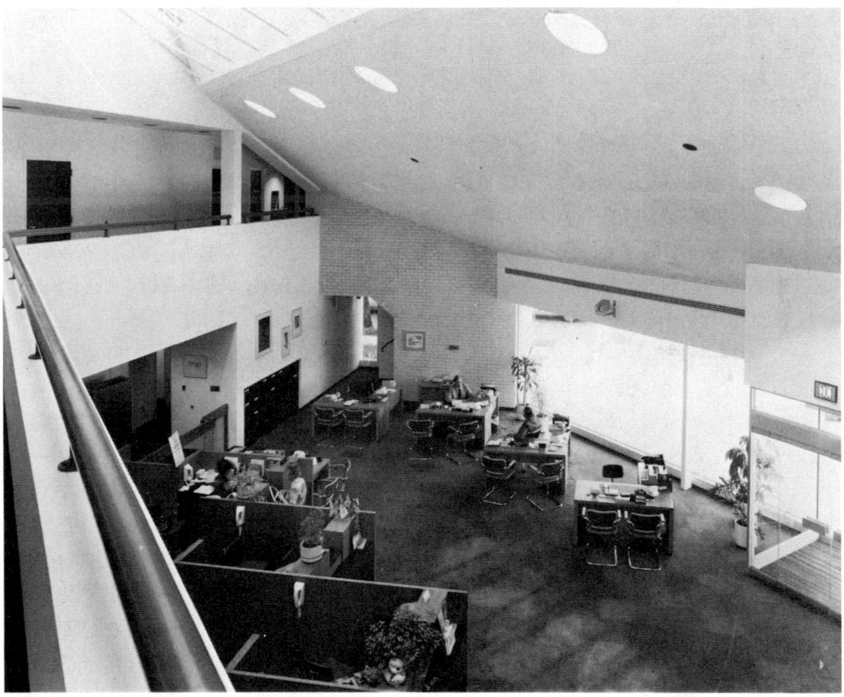

Figure 9–5. Tellers' area and office space as seen from the mezzanine passageway above.

Figure 9–6. The main banking and office area looking into the public space. The mezzanine area, above and to the right, houses office space.

super metal halide lamp at 100 lumens/watt is, of course, still more efficient but at the time the project was in the design stage, this lamp was suitable for only horizontal mounting. Since downlight fixtures were selected for aesthetic reasons, the vertical burning 400-watt MH lamp was the obvious choice. (See Figure 9–7.)

For office lighting, 175- and 250-watt MV fixtures are generally used. At the lower ceiling heights, higher wattage lamps would have been too bright and would produce too much glare. Despite the higher efficacies of the phosphor coated MH lamps, which are available in those wattages, MV lamps are used. (See Figure 9–8.) This choice was based on an analysis of life-cycle cost for the two lamp types. The MH lamps, although more efficient than the MV, are not long-lived in the 175- and 250-watt sizes and are much more costly than the MV lamp. Thus, life-cycle costs due to lamp replacement using MH lamps for office lighting would have been higher. This, of course, would have been partially offset by the savings of energy made possible by using fewer MH fixtures. However, at the conventional 9-foot ceiling height in the office area, the use of fewer fixtures offered no advantge since they

Figure 9–7. Lamp data. (Lamp data, © 1977, Gardco Manufacturing Inc. Author's amendations are marked by *.)

LAMP TYPE		WATTS (Nominal)	LAMP ORDERING NUMBER	INITIAL LUMENS	LUMENS PER WATT	RATED LIFE
METAL HALIDE	CLEAR HORIZONTAL	1500	M1500/BU-HOR	150,000	100.	1,500
		1000	G.E. MV1000/U	107,800	107.8	10,000
		1000	M1000/BU-HOR	95,000	95.	10,000
		400	SYL. MS400/HOR	40,000	100.	*20,000
		400	M400/BU-HOR	*34,000	*85.	*20,000
		250	M250/BU-HOR	19,500	78.	*10,000
		175	SYL. MS175/HOR	15,000	85.7	7,500
	CLEAR VERTICAL	1000	SYL. MS1000/BU	125,000	125.	10,000
		1000	G.E. MV1000/BU/1	115,000	115.	10,000
		1000	West. MH1000/BU	100,000	100.	10,000
		400	M400/BU	34,000	85.	15,000
HIGH PRESSURE SODIUM		1000	LU1000	140,000	140.	24,000
		400	LU400	50,000	125.	24,000
		310	G.E. LU310	37,000	119.4	24,000
		250	G.E. LU250/BU/S	30,000	120.	24,000
		250	LU250	25,500	102.	24,000
		150	LU150/55	16,000	106.7	24,000
		150	LU150/100	16,000	106.7	24,000
		100	LU100	9,500	95.	*24,000
		70	LU70	5,800	82.9	*24,000
DELUXE WHITE MERCURY		1000	H1000DX36-15	60,000	60.	24,000
		400	H400DX33-1	21,750	54.4	24,000
		250	H250DX37-5	12,300	49.2	24,000
		175	H175DX39-22	8,180	46.7	24,000
		100	H100DX38-4	4,200	42.0	24,000
*MET HAL. CLR. VERT		400	SYL. MS400	40,000	100	15,000

UPDATING FOOTCANDLE VALUES

$$\frac{\text{New Lamp Lumen Output}}{\text{Old Lamp Lumen Output}} = \frac{\text{Updating}}{\text{Multiplier}} \qquad \text{Example: 400 W Metal Halide } \frac{40000}{32000} = 1.25 \text{ Factor}$$

Old Footcandle Value x Updating Factor = New Footcandle Value

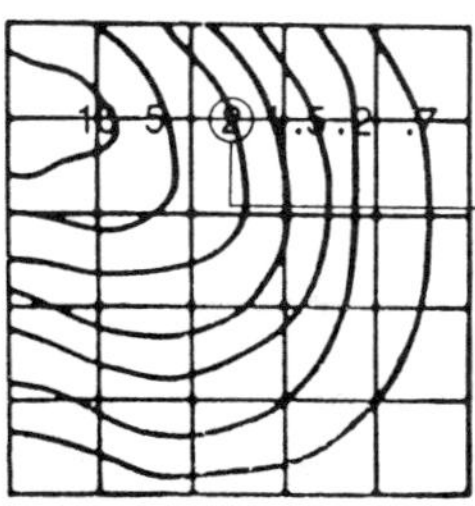

Example:

2FC x 1.25 = 2.5 Initial Footcandles (Based on M400 rated at 40000 lumens).

EXISTING FOOTCANDLES
(Based on M400
Rated at 32000 Lumens).

would have exceeded their designed spacing-to-mounting-height ratio. For optimum lighting, a fixture S/MH ratio of 1 to 1.5 through 1 to 1.8 is the maximum desirable for these particular downlights selected for this project. Therefore, MV fixtures were the choice for these applications since the design spacing layout would not exceed their published S/MH ratio. The offices are illuminated to 70 fc.

LAMP EFFICACIES					
LAMP TYPE AND WATTAGE	LAMP COST*	PRESENT LAMP COST	LAMP LIFE (HOURS)	INITAL LUMENS	EFFICACY LUMENS/WATT
M.V. 175	$14.00	$16.50	24,000	8,500	49
M.H. 175	41.00	49.00	7,500	14,000	80
M.V. 250	22.75	27.25	24,000	13,000	52
M.H. 250	44.50	53.00	10,000	20,500	82

* AT TIME OF DECISION ON DESIGN

LAMP EFFECTIVENESS COMPARING METAL HALIDE (MH) AND MERCURY VAPOR (MV) LAMPS.

Figure 9–8. Comparing mercury vapor and metal halide lamps.

For accent lighting in several locations, 100-watt, MV, up-down light fixtures are used, each with a dual purpose. At the exposed brick stairwell, the fixtures provide both a low level of safety lighting and a soft wash of light to graze and highlight the texture of the brick. The same fixtures mounted near the building skylight (see Figures 9–3 and

9–6) serve both to illuminate the mezzanine passageway and at night to uplight the skylight from within, becoming a part of the building's exterior lighting. (See Figure 9–4.)

Restroom lighting is by 100-watt MV fixtures, with one downlight fixture mounted in each ceiling. In addition, decorative MV fixtures are mounted in the glass of each restroom mirror. The fixtures, each consisting of a chrome lampholder and G-40 lamp, are fed from remote ballasts located in the dropped ceiling cavity. (See Figure 9–9.)

In order to prevent accidental extinguishing of the lamps, all HID lamp interior lighting is controlled by key-operated switches. If the

Figure 9–9. New decorative vanity lighting utilizes a different, innovative approach. 100W, G-40 MV lamps are installed in chrome lampholders mounted within the vanity mirror, producing a different image, and is a very functional utilization concept.

MV or MH lamps are turned off during working hours, they would require from 10 to 15 minutes of cooling time before the arc could restrike, and several more minutes to reach full intensity. For the same reason, an inverter/rectifier emergency power unit is provided to maintain power to the emergency white HID fixtures during a power outage. Exterior lighting at the building soffit, drive-up banking locations and rear wall adjacent to the parking area is by 175-watt MV fixtures. (See Figures 9–1 and 9–2.) These lighting fixtures are time-clock controlled in conjucntion with a photocell.

The overall design achieved a successful blend of aesthetics, energy management and energy conservation, with an evenness of illumination and color rendition equal to fluorescent lighting in a more conventional building.

AMERICAN FEDERAL SAVINGS & LOAN, WASHINGTON, DC

Like NS&T, American Federal Savings & Loan, also in Washington, DC, winner of the IES Edwin F. Guth Lighting Design Award of Merit, is a building of unusual shape—a combination of circular and rectangular elements. The main banking area, second floor office and basement storage areas comprise the round structure. The restrooms, lounge, stairs and a small storage area are in the rectangular portion. (See Figures 9–10 and 9–11.) As with NS&T, this irregular configuration argued against the use of conventional fluorescent lighting.

The electrical (artificial) illumination system which was installed uses MV lighting extensively. The main banking and office areas are lighted with circular patterns of 175-watt, MV, downlight fixtures to 70 fc, while the tellers' area has 250-watt MV fixtures to provide 150 fc. (See Figures 9–12, 9–13 and 9–14.) Color-corrected lamps, rather than Brite White Deluxe lamps, are used in the tellers' area, offices and public space to create warm tones which will accent the wood cabinetry.

Because of architectural considerations, the architectural partner in charge originally decided on incandescent downlights which have an extremely low efficacy of only 17 to 18 lumens/watt. This would have required three times more electric energy than the MV lamps used (which produce 40 to 52 lumens/watt for the lamp wattage range sel-

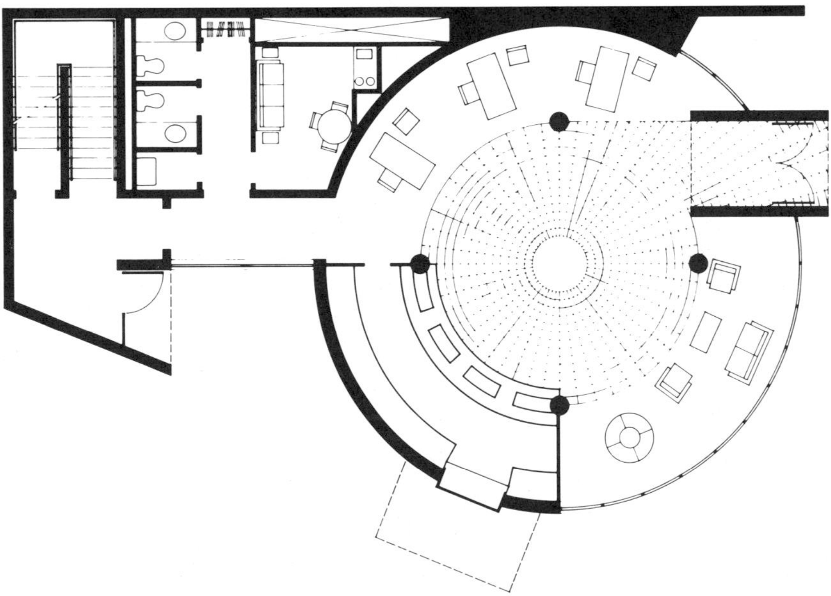

Figure 9–10. Floor plan of American Federal, first floor.

ected for this project). In addition, three times more electric energy would have been needed for air conditioning to remove the heat generated by the incandescent lamps. The IES recommended minimum illumination was 70 fc in offices and public space, and 150 fc in tellers' area. A compromise was made by using energy efficient 175- and 250-watt, color-corrected, mercury vapor lamps in milligrove (micro-baffle), high hat (downlight) fixtures, which also reduce aperture brightness.

Globe-type MV fixtures are used in the basement storage areas, again in a circular pattern, and 100- and 175-watt MV downlights are used throughout corridors, stairs, restrooms and the employee lounge to provide a lighting installation of uniform attractiveness. (See Figures 9–15 and 9–16.)

Two special aspects of the lighting design are the main lobby chandelier and the parking area lighting. The main lobby fixture is an outdoor roadway luminaire featuring a bronze-tinted drum approximately 28 inches in diameter by 11 inches deep, intended for pole mounting; the fixture (without pole) is instead inverted and hung from the ceiling. Because of its relatively short tenon, the fixture appears to

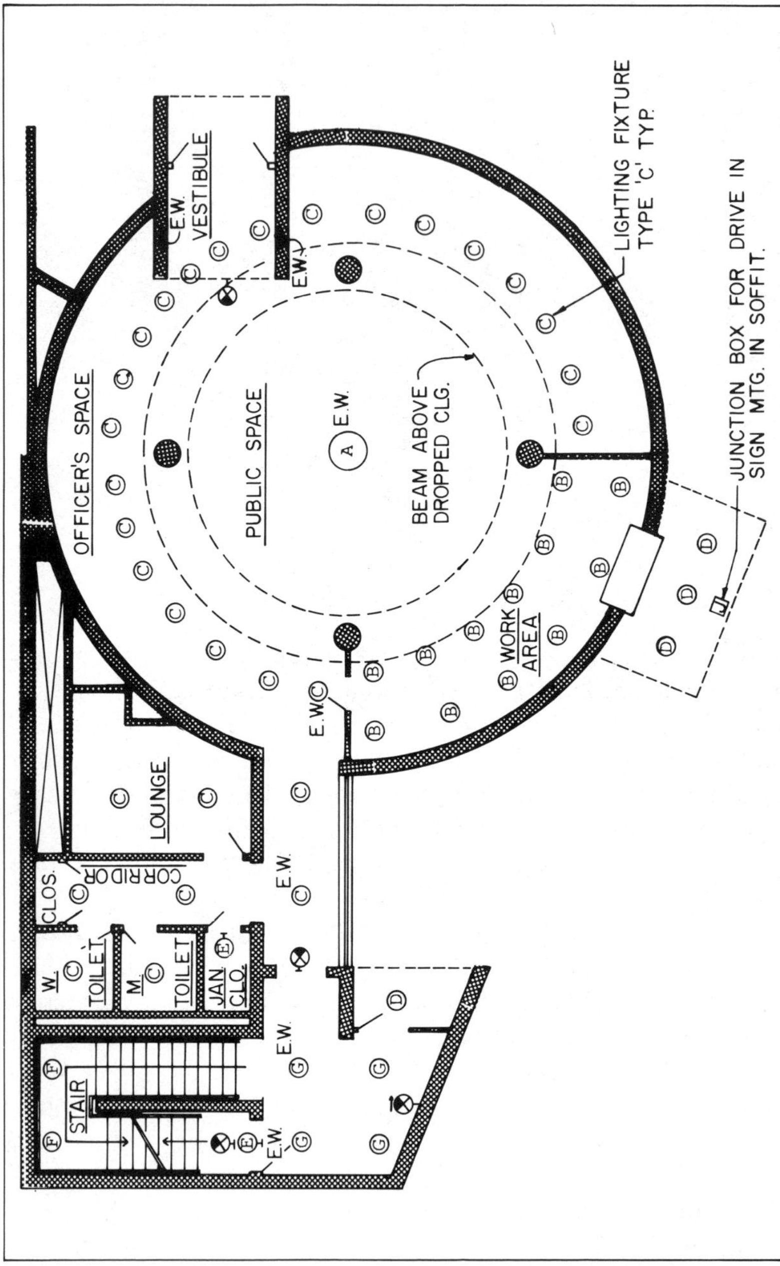

Figure 9–11. Banks first floor lighting layout.

Figure 9–12. The tellers' area is to the left and the officers' space is straight ahead and slightly to the right as seen from this outside view.

Figure 9–13. The same view as Figure 9–12, taken from inside of the main banking and office area. The overhead outdoor chandelier roadway luminaire compliments the circular pattern of tiles in the public space, as well as the centrally located customer service desk. 175-watt mercury vapor downlights are used in a circular pattern.

Figure 9–14. Public space and tellers' area with the author contemplating a transaction.

hover in the center of the main banking area about 10 inches below the ceiling. (See Figures 9–12, 9–13, 9–17, 9–18 and 9–19.) It is an indirect fixture equipped with a top (or in this application bottom) aluminum reflector and uses a single 400-watt, R-60, non-phosphor-coated, MV lamp. The face of this lamp does not have a phosphor coating to color-correct the spectral power distribution since this would destroy the light control photometrics of the lamp. The color of the light transmitted is improved and corrected by using a bronze-tinted drum lens. Although some loss of efficiency is encountered, the color of light is enhanced and is very compatible with the warm tones of the color-corrected MV lamps used in the other fixtures.

The fixture selected to illuminate the parking area was chosen for its architectural as well as its photometric qualities. The wall of the American Federal Building facing the parking lot is an unbroken rectangle of exposed concrete with a pronounced pattern of control joints and bolt holes. Thus, the fixture used is a wall packet of perfectly square design and is mounted in the geometric center of the wall, 15 feet above grade. It looks as good by day as by night and uses a 400-watt SMH lamp. Other exterior lighting at the drive-up banking

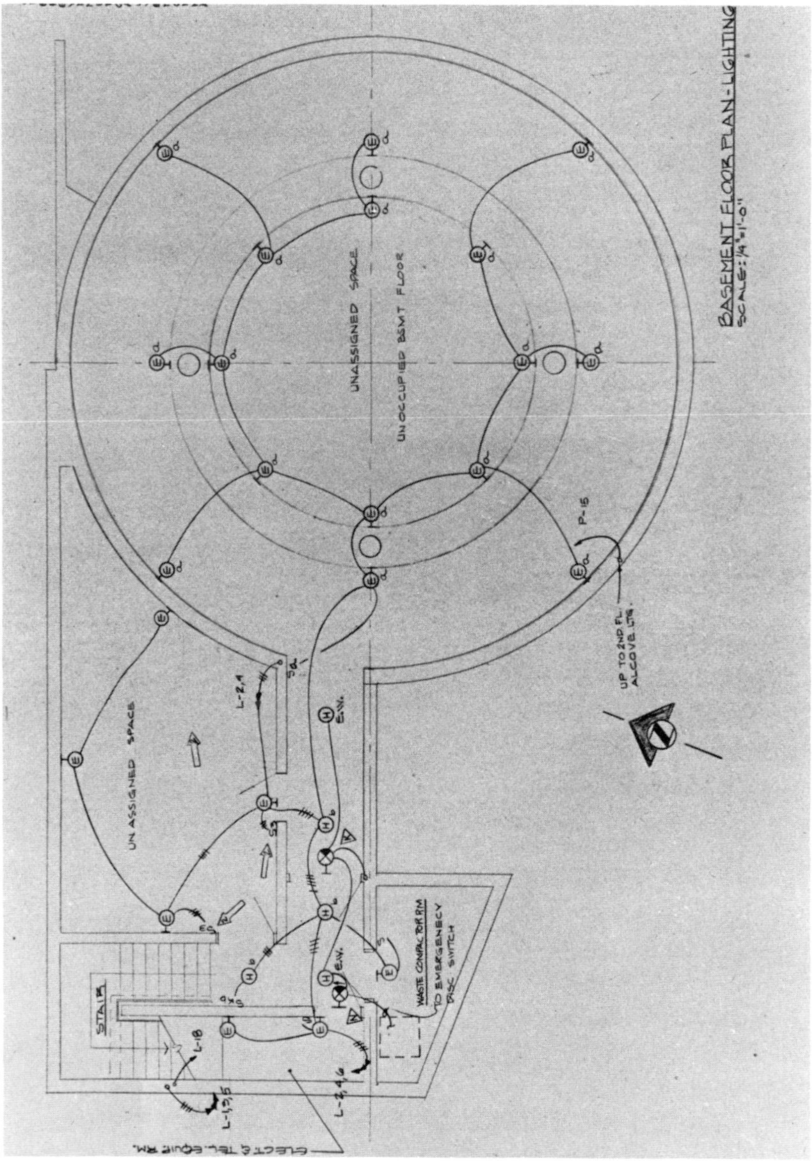

Figure 9–15. Schematic basement floor plan-lighting design layout.

Figure 9–16. Wall (beam) bracket globe type 175W mercury vapor lighting fixtures are utilized to illuminate the basement storage area.

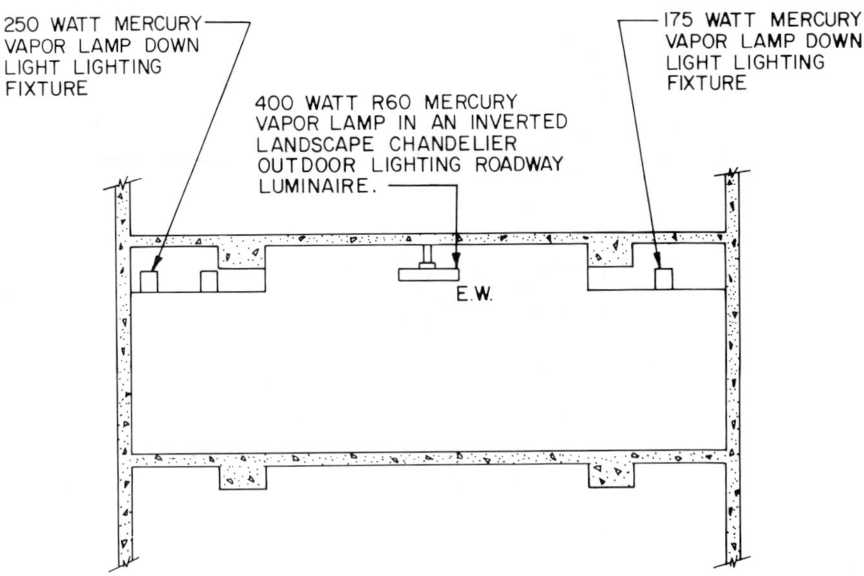

Figure 9–17. First floor banking lighting layout (sectional view).

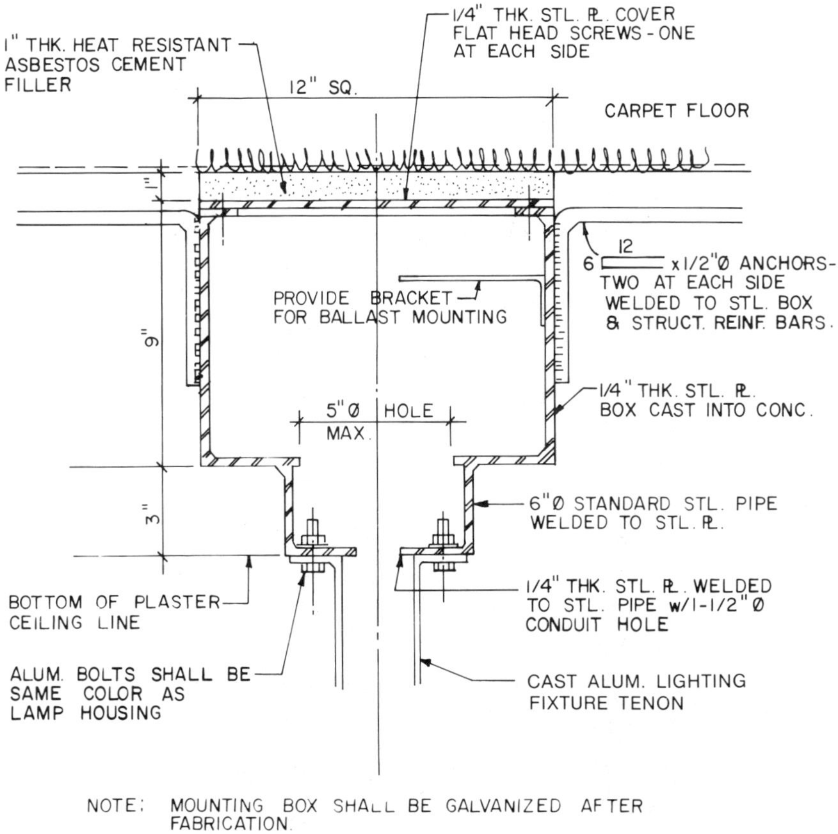

Figure 9-18. Ballast box detail for light fixture mounting.

station and rear door uses 175-watt, MV, downlight fixtures. (See Figure 9-20.)

Voltage Used for Lighting

All lighting at American Federal S&L operates at 277 volts, as does some of the mechanical equipment, which also operates at 480 V. It is, of course, desirable to serve as much of the building load as possible at 277/480 V, since this voltge can carry 231% as much energy as a 120/208 V system. Providing lighting at 277 V also reduces the num-

Figure 9–19. The main lobby of AFS&L has an outdoor roadway luminaire which is used as a chandelier. This fixture uses a single 400-W, R-60, nonphosphorcoated MV lamp, color corrected by a bronze tinted drum lens.

ber of circuit breakers required to serve a given load, reduces the size of the lighting panel's feeder, reduces the panelboard's bus ampacity requirements and reduces the size of the feeder circuit breaker in the main distribution equipment. Use of 277/480 V, wherever possible, reduces the first cost of the overall electrical distribution system and reduces the life-cycle cost, because such an electrical design can incorporate smaller and fewer pieces of electrical equipment requiring servicing. Lighting at NS&T operates at 120 V because of the existing availability of 120/208 V utility secondary power at the site.

Figure 9–20. A wall mounted square wallpacket utilizing a 400-W super metal halide lamp illuminates the parking area and the architectural features of AFS&L's concrete wall. In the background, several 175-W, MV downlights illuminate the drive-up banking station and rear door.

MARTIN MARIETTA CORPORATE HEADQUARTERS, ROCKVILLE, MARYLAND

The entrance of the corporate headquarters of Martin Marietta properties is both pleasing and inviting. The main entrance lobby of this $18 million complex is illuminated to 50 fc by four overhead, 400-watt, MH downlights that are 45 feet above floor level. The walls form a triangular opening, like a funnel, with ever-increasing levels of illumination as the building's wings come together, leading a visitor to the elevator lobby. (See Figure 9–21.) All interior lighting operates at 277 volts in order to reduce line losses in the form of heat, reduce voltage drop and decrease materials used for construction, which reduces first costs, maintenance costs and life-cycle costs.

The cove lighting on all five levels of the elevator lobby utilizes two

Figure 9–21. Martin Marietta corporate headquarters. Increasing levels of illumination, as the building's wings come together, lead a visitor to the lobby.

40-watt, rapid-start, cool white fluorescent lamps behind a low glare acrylic lens. The lens reduces glare for those few who might look directly up into the fixture, and it also minimizes hot-spotting on the wall adjacent to the fixture. The cove fixture has its ballast adjacent to the wall and lamps nearest the center of the lobby. This technique causes a high grazing angle and accents the rough precast wall panels. Both high VCP qualities and good, soft ESI are obtained in this area. (See Figure 9–22.)

The first area most visitors see is the parking garage. This is illuminated to 20 fc using two U-shaped 40-watt lamps in 2′ × 2′ fluorescent fixtures regressed into the concrete ceiling pans as shown in Figure 9–23. The fixtures appear as floating rectangles of light because of their low brightness which is created by using high VCP lenses. This effect is much more striking than ordinary utilitarian garage lighting and very appropriate for a major corporate headquarters. (See Figure 9–24.)

The basic lighting fixture for the office building is a 3′ × 4′ 4-lamp, 40-watt fluorescent fixture with a 3′ × 3′ square lens. This is used as part of a coffered 5-foot modular integrated ceiling system that includes an air supply through linear air bars. Since lampholders and the normally darker lamp ends are concealed behind the lamp door frame, the fixture has a particularly uniformly lighted appearance. At a 9-foot ceiling height, the fixtures produce 90 fc at a visual comfort index of 88%.

The fixtures are the dynamic air handling heat-return type. By using

Figure 9-22. Cove lighting provides an effective technique that accents the rough precast wall panels of the hallways.

this *heat-of-light* concept, the efficacy of the fluorescent lamps is increased by about 25%, delivering more lumens per watt. Also, lamp and ballast life are increased, and the heat load injected into the working space is reduced to the minimum. Furthermore, returning fixture heat directly to the ceiling plenum results in a high temperature differential for a relatively small quantity of air, which increases the efficiency of the cooling system.

In the cafeteria area, 100-watt MV Brite White Deluxe lamps, installed in milligroove downlights, are used in the building's standard 5′ × 5′ coffered ceiling to obtain a comfortable 45 fc level. (See Figure 9-25.) The cafeteria lounge area utilizes the same lighting fixtures and design approach. Using illumination to visually separate the large open space of the cafeteria lounge from the corridor, 100-watt MV downlights, mounted on 5-foot centers are installed in a cove adjacent to the corridor's soffit, creating a scalloping effect. (See Figure 9-26.) The eveness of the illumination that is produced and the separation effect are shown in Figures 9-26 and 9-27.

Figure 9–23. Illumination of the parking area is maintained at 20 fc using two U-shaped, 40-W lamps in 2′ by 2′ flourescent fixtures.

Figure 9–24. The evenness of the overall garage lighting can be appreciated in this view. Both high illumination levels of good quality lighting are abundant throughout this facility as well as in between the cars. The cars are readily identifiable by color, and the door key slots are well lit.

Figure 9–25. The cafeteria is illuminated to 45 FC of pleasing light having good color rendering properties by the use of one 100W bright white deluxe MV lamp downlight fixture per each 5′ × 5′ coffered ceiling pattern.

Figure 9–26. The cafeteria lounge area utilizes the same lighting fixtures as the cafeteria. The lounge area is visually separated from the public space and corridor to the left by creating a scalloping effect on the wall which acts as a curtain created by illumination. 4.8 watts/ft^2 are used to create this striking effect.

Figure 9–27. Separation of the lounge area from the corridor is done by lighting. A comfortable 45 fc is maintained in the lounge with 100-W MV downlights, on 5-ft centers, thus forming a lighting curtain that helps to accent and visually emphasize the corridor while separating it from the lounge.

Restroom lighting is similar to that at NS&T. Figures 9–28 and 9–29.) illustrate a wiring and lighting solution for this facility's toilets. The decorative wash basin luminaires use a G-40, deluxe-white, MV lamp in a chrome plated lampholder. The lampholder is fed from an overhead, field-fabricated, step-down transformer and 100-watt, high-power-factor, constant-wattage, auto-transformer ballast mounted in a NEMA-1 enclosure located within the dropped ceiling cavity space. As all lighting at Martin Marietta is served at 277 volts and the lampholders are mounted at only 6 feet 6 inches above the finished floor, the transformer was required to step-down the 277-volt power to serve the lampholders and lamps at 120 volts. This is a National Electrical Code requirement, Article 210–6 (voltage to screw-shell lampholders mounted below 8 feet A.F.F. shall not exceed 150 volts to ground). This solution not only permitted single point switching of the toilet rooms' overhead fluorescent and wall-mounted, vanity-mirror, MV fixtures, but eliminated the need for 120-volt fixtures

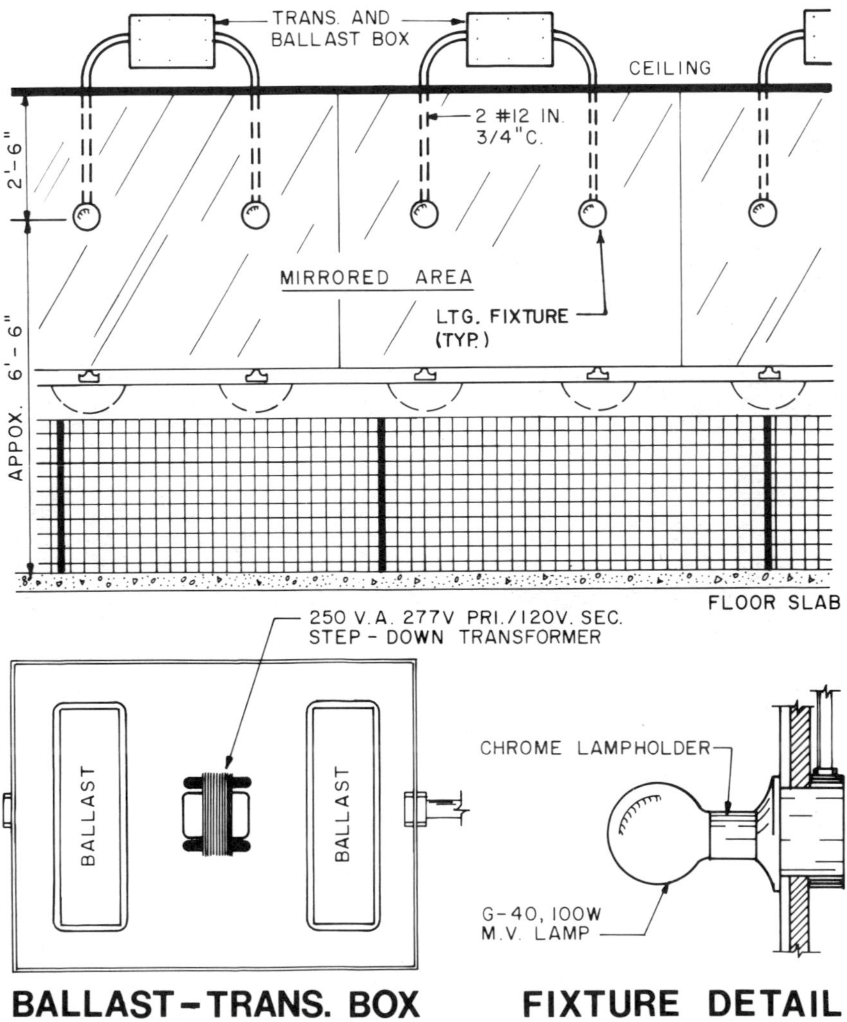

Figure 9–28. Decorative lighting utilizing a 120-volt screw shell lampholder.

and circuitry on a predominantly 277-volt project. This concept alleviated the need to abandon a successful lighting design because of potentially troublesome NEC restrictions.

This lighting design and approach allowed a custom design with standard off-the-shelf products having very high efficacies, long life and low first cost, as well as reduced energy and life-cycle costs.

Figure 9–29. Aesthetic decorative vanity lighting fixtures are mounted in the wall-to-wall mirror.

DIAMOND ELEMENTARY SCHOOL, MONTGOMERY COUNTY, MARYLAND

All lighting operates at 277 volts and incorporates the heat-of-light principle in its use of dynamic air handling fixtures. By returning air through the ends of the recessed gird troffers, and up through the lamp and ballast compartments, the efficacy of the fluorescent lamps was increased by 25% and lamp and ballast life were substantially increased. (See Figure 9–30.)

Fluorescent lamps operate most efficiently at a certain internal mercury vapor pressure. If this pressure either increases due to lamp overheating (as in static, non-air handling enclosed fixtures) or decreases due to cooling (as in outdoor fixtures or dynamic fixtures which supply air through the lamp compartment), lamp light output as measured in lumens/watt decreases.

Returning air through properly designed dynamic air handling fixtures maintains optimum lamp mercury vapor pressure and maximizes

Figure 9–30. Dynamic air handling grid troffer lighting fixtures utilize an attractive black matte around their regressed lens frame holder which houses a KSH K-19T (silvertin lens). This is the IMF room in the Diamond Elementary School.

light output. Maintenance of this optimal temperature within the fixture also serves to extend lamp and ballast life.

Just as importantly, the use of dynamic heat return air-handling fixtures reduces the heat introduced into the working space. At 45 CFM air return through the fixture ends, a quantity which is generally not exceeded because of static pressure limitations, approximately 70% of the fixtures' heat is returned directly to the ceiling plenum. This, in turn, reduces the air conditioning load of the working space and reduces the size of ductwork and air conditioning equipment required for the space, since in general, one watt of A.C. load is needed to remove every two watts of lighting heat load. (See Figure 9–31.)

The use of standard, cool-white-series fluorescent lamps offered nearly perfect interior color quality, and emphasized the natural tone of the wood cabinetry used throughout the school by including the richer texture which would have been absent if another light source

Figure 9–31. 120 FC's of field measured illumination having 70 ESI of FC's are present in this classroom while consuming less than 3 watts per square foot.

had been utilized. The classrooms, in general, are illuminated to 120 field-measured footcandles having over 70 footcandles of equivalent sphere illumination while consuming slightly less than three watts per square foot. What could make it apper as a blatant contradiction is that a visual comfort probability of nearly 80 was also maintained, a seeming impossibility when one considers that ESI normally utilizes fixtures and lenses that distribute light in a batwing primarily in the 65°, 75° and 85° zones, while VCP lighting fixtures have calculations based on restricting the light distribution in the 65°, 75° and 85° zones. (See Figure 9–32.)

In the gymnasium, 60 footcandles are maintained in an area of 3,500 square feet with a ceiling height of 20 feet, with 12–100% downlight, totally enclosed, industrial, 400-watt, MH lighting fixtures offering 85 lumens per watt and 15,000 hours of life (now 20,000 hours). The even illumination can be attested to by the light scalloping on the walls at a height of approximately 12 feet. This even illumination and overlay at this height will prevent the apparent time lag or

Figure 9–32. The IMF room has a clear, modern exit sign in front of us and to our right that clearly delineates the egress routes. The emergency fluorescent lighting fixtures produce 10 FC's at all egress routes and points of departure.

stroboscopic effect which can be a problem in active sports such as basketball and volleyball. Various lighting systems were used and controlled to allow separate or combined use of fluorescent and metal halide lighting fixtures in order to tailor illumination levels for either gymnastics, volleyball, basketball, community sports, or community parties and dances. By shutting off the key-operated MH lighting fixtures, for instance, we are left with 10 fc of fluorescent lighting. Energy use is 1.63 watts/ft^2 or 37 fc/watt with the MH lighting system. (See Figure 9–33.)

The fluorescent fixtures on the emergency circuit have wire guards for physical protection and maintain 5 footcandles of minimum illumination. All emergency white lighting fixtures are maintained throughout an emergency by the use of OEM integrally installed, standby power packages within their ballast channels.

Throughout this facility, all lighting was designed as an integral part of a total concept in order to be functional and still architectur-

Figure 9–33. The gymnasium at Diamond Elementary School has both M.H. lighting sources and fluorescent lighting for emergency standby by using O.E.M. installed inverter/rectifiers.

ally aesthetic. The multipurpose room uses one of the first convenient, continuous, full-range, 1000-watt, MV dimming systems in the country, accommodating a range of activities, from dining to plays to dances. The lighting can be immediately changed in intensity, up or down, by 70%. The remaining 30% of intensity change follows the warm-up and cool-down characteristics of the HID lamps and takes another 3 to 5 minutes. The system was deliberately designed to 120 fc though normally we need only a maximum of 70 fc maintained. By having higher levels of illumination than required, we can maintain 70 fc even to the end of lamp life, thus substantially reducing maintenance and material cost. The stage has its own separate 1000-watt MV dimming system.

The multipurpose room and stage area have 2′ × 2′ recessed grid fixtures with black matte reveal trim and recessed lenses to complement the tone and style set throughout this facility. (See Figure 9–34.) One of the major aspects of this area is that when less light is needed and used, then less electric energy is used in direct proportion. A 50% illumination level will consume 50% of the electric energy as was used

Figure 9–34. This is the multipurpose room that doubles as a cafeteria and stage play production area, but also has such other diversified activities as community parties and dances.

at the 100% lighting level. The overall electrical design was low in initial cost (less than 9% of the overall cost) and is proving low in operating cost.

DUAL PURPOSE CONFERENCE ROOM AND TELEVISION STUDIO (PICTUREPHONE): CHESAPEAKE AND POTOMAC TELEPHONE COMPANY, WASHINGTON, DC

The Picturephone Facility lighting is a combination of dimmed MV and fluorescent fixtures, giving a wide range of possible lighting levels.

In the seating area, 100-watt MV downlights at their highest setting produce 40 fc, while in the lobby, 175-watt MV downlights were used along with dimmer-controlled fluorescent cove lighting. (See Figure 9–35.) From the wall back into the center of the room, illumination decreases from 70 fc to 50 fc. (See Figure 9–36.) The good quality of the cool white fluorescent lamps and the white deluxe mercury vapor

Figure 9–35. The receptionist desk is to the left, the coffee bar to the right and the lobby waiting area is in the center. This central waiting area is decoratively and attractively illuminated by five 100-W MV downlights that are dimmable by solid state dimmer controllers.

lamps gives true color rendering to the vinyl wallpaper collage. (See Figure 9–37.) All dimming equipment used is solid-state, representing the latest technological state of the art, having long life and maintaining quiet and efficient control of energy and light. (See Figure 9–38.)

Figure 9–36. The dimmable fluorescent cove lighting in this lobby area accents the collage wall covering to create a comfortable, attractive space. In addition, the recessed display shelf and accent lighting can be seen.

Figure 9–37. Here is a close-up of the recessed telephone display shelf and accent lighting that is both functional and aesthetically pleasing. The receptionist desk is to the far right.

Figure 9–38. This is the nighttime view of the lobby office. Fluorescent cove lighting and 175-W mercury vapor downlights provide illumination. Both are dimmer controlled which has the additional advantage of reducing energy consumption and extending the lamp life while reducing life cycle costs.

RENOVATION: RICHARD MONTGOMERY HIGH SCHOOL, ROCKVILLE, MARYLAND

Richard Montgomery High School is a complex of buildings and additions constructed over more than a 40-year period, and recently rehabilitated at a cost of $3.2 million. The completed renovation encompasses 240,000 square feet of space, including gymnasiums, multipurpose room, ballet practice room, library, cafeteria, TV studio and classrooms.

The challenge facing the engineer in the course of this major rehabilitation was the upgrading of a dated and energy-inefficient facility into conformance with modern technological breakthroughs and innovations in the field of energy conservation. As a measure of success it should be pointed out that although the design of this facility was completed more than seven years ago, it is still in compliance with the current BOCA Energy Conservation Code and ASHRAE 90–75.

The master gymnasium, where all intramural sports are played, can be divided into two separate gyms with independent lighting and sound systems. It is lighted with 20–100% downlight, enclosed, industrial, 400-watt MH fixtures offering 85 lumens/watt at 20,000 hours' life, and produces 62 fc when divided and 80 fc when opened for joint use. In the divided mode, we obtain 56 fc per watt and in the open mode 72 fc are produced per watt, at an energy use of 1.11 watts per square foot. (See Figure 9–39.) As at Diamond Elementary School, fluorescent lighting is used in conjunction with the MV lamp for better color quality and inexpensive emergency lighting. The fluorescent lighting acts as a standby lighting source in the event a high speed utility strike extinguishes the HID arc plasma which would require a 10 to 14 minute restrike time. Thus, the emergency lighting system can be maintained throughout an emergency by a standby emergency generator rather than a more sophisticated and expensive uninterruptable power system. The fluorescent fixtures maintain 5 fc of minimum illumination and are provided with wire guards for physical protection.

In gymnasium number 2, a maintained 70 footcandles is established in an area of 5,400 square feet with a ceiling height of 22 feet by the use of 15 of the same MH fixtures, and again the even illumination at even

Figure 9–39. This master gymnasium is where all intramural sports are played at Richard Montgomery High School. It can be partitioned into two separate gyms, each with its own independent lighting and sound system.

Figure 9–40. In gymnasium No. 2 the even illumination at the 12-foot high level can prevent the apparent time lag or synchronization phase shift due to the stroboscopic effect when a ball passes from one light source through darkness and into light again, as can happen in such sports as volleyball or basketball.

a 12-foot height serves to reduce the strobe effect on moving objects. Energy consumption in gym No. 2 is 1.31 watt/ft^2 or 53.4 fc/watt. (See Figure 9–40.)

The exercise room utilizes the first continuous, full-range dimming, SMH lighting system designed for use in this country, although there are other similar systems that became operational before this facility. (See Figure 9–41.) There is no apparent color shift as the intensity of illumination is brought up. (See Figure 9–42.) The lighting can be immediately changed in intensity, up or down, by 70%. The remaining 30% of intensity change follows the typical warm up or cool down characteristics of a HID lamp and takes approximately 3 to 5 minutes. Energy consumption varies in proportion to the lighting level; thus, the system was deliberately designed to deliver about 100 footcandles even though in most conditions we need a maximum of only 70 to 80 footcandles maintained.

The reason for this apparent design discrepancy is that HID sources, such as MV, SMH and HPS lamps, are rated at initial lighting output. If the maintenance factor were not considered when establishing a maintained illumination level, a lighting installation designed for

Figure 9–41. This exercise room is presently at its highest illumination level of 100 FC's. It utilizes a manual energy and lighting control device to control or regulate its illumination level.

Figure 9–42. Here the author adjusts the manual energy and lighting controller panel in selecting a function requiring a lower illumination level. There are three function selections and illumination levels available to participants using the room. The highest level is 100 FC, although 60 FC is all that is normally required. The exercising function selects 60 FC's and aerobic dancing selects 30 FC's. The user selects the function that the room is going to be used for and does not know what illumination level he is selecting. In this view 30 FC's are utilized.

a particular illumination level would start at that value only initially; the illumination would then steadily decrease from the day of installation. When using lamps and lighting fixtures rated at 0.7 maintenance factor, a typical lighting system intended to deliver 70 fc is designed to deliver an initial 100 fc. Over the life of the lamps such a system will provide an average 70 fc but will provide an actual 70 fc for only part of its life. However, the energy consumed over the entire period of usage is 100% of the connected load, whether it produces 100 fc or 70 fc. With a dimming system such as that used in the exercise room, lighting levels can be maintained at the desired designed intensity with subsequent savings in energy use and increased lamp life.

The multipurpose room's dimming system utilizes a photocell override that provides feedback data to an automatic energy control system, to accommodate such diversified activities and illumination levels as: exhibitions and matches at 70 fc general exercising and recreation at 30 fc (see Figure 9–43) assemblies at 20 fc and dances at 5 fc. (See Figure 9–44.) The photocell works in conjunction with a

Figure 9–43. This AEC system controller can accommodate such functions as exhibitions and matches, as well as general exercising and recreation, assemblies or dances. Just like the exercise room, the occupant selects the activity to be engaged in by dialing it upon the AEC system control panel. The individual does not per se select the illumination level.

Figure 9–44. The Richard Montgomery High School multipurpose room utilizes one of the first automatic energy control (AEC) systems in conjunction with a photocell for feedback control of the AEC system.

preset dimming control. When a given preselected illumination level, such as 20 fc for an assembly, is reached, the photocell holds it at that intensity thereby not utilizing any additional and unnecessary lighting or energy.

Using the AEC system, which is available from several manufacturers, an initial illumination level of 115 fc was designed in order to maintain 70 fc during the life cycle. Only 60% of the energy is utilized initially to produce 60% of the initial illumination, or 70 fc. As the system ages, more energy is required until many thousands of hours later, nearly full 100% energy usage is required to produce 70 fc maintained. Thus, the energy saved over the life of the lamp represents a substantial cost saving while maintaining the desired 70 fc throughout the installation.

With AEC the HID lamps can be utilized to the end of their life cycle of 20,000 hours for SMH, while a conventional SMH lighting system would be relamped at 70% of maintained lumens, or approximately 15,000 hours. The AEC-controlled, HID dimming system reduces the maintenance cost attributable to more frequent labor and material changes required by a conventional non-dimmed HID lighting system, and makes group relamping possible, which is more economical than spot relamping frequently necessary with a conventional non-dimmed system.

Thus, by controlling dimmable MH luminaires with an AEC system, four levels of illumination can be realized for various activities, and by operating the system in a dimmed mode at an average 70 fc rather than at full intensity (or 115 fc initially available), the energy savings alone, based on the then current 4¢ per kilowatt hour, was $198.54 per year or $808.90 between lamp changes, based on the original 15,000-hour life. Since the system cost without dimming was $2,400 and with AEC was $3,200, this meant that with AEC the energy savings alone per lamp change made up the cost difference within the first lamp change.

Maintenance costs were based upon $43.55 per lamp and $10.00 labor per fixture for cleaning and relamping. Within a 9.6 year period, which was the projected life expectancy for this facility, $695 was saved in material and another $160.00 was saved in labor, for a total savings in maintenance changes of $855.00. With energy savings of $1,920.00 over a 9.6 year period, the total savings from using AEC would be $2,775.00.

Richard Montgomery's library is illuminated with fluorescent fixtures, giving a pleasing natural tone to the wood cabinetry. (See Figure 9–45.) The illumination is a field-measured 120 fc at slightly less than 2.9 watts per square foot power consumption. A visual comfort probability of over 80 was maintained with 4-lamp, 40-watt, 2′ × 4′ recessed fixtures which restrict light distribution in the 65° to 85° zone of offending glare.

The entrance passageway to the library is lighted to a lower level to set the lobby apart from the main corridor. (See Figure 9–46.) 100-watt recessed downlights with Deluxe Bright White MV lamps were used, primarily for mood effect, and produce 30 fc. The fixtures were specified with black milligroove baffles to reduce aperture brightness.

Typical classroom lighting uses alternate ballast switching. One switch controls the ballast of the inboard tubes of the 2′ × 4′ fluorescent fixtures, while the second switch controls the ballast of the outboard tubes. This provides the option of having all the lamps on to give a high, even illumination of up to 100 fc. (See Figure 9–47.) If this illumination is too high for a given function, half of the tubes in each

Figure 9–45. The renovation and addition leading to this startling new library facility is illuminated to 120 field measured FC's. Good quality lighting was in part attributable to a full color spectrum of illumination produced by the cool white fluorescent lamps and silver tinted male conic lenses used.

Figure 9–46. The entranceway to the library is illuminated to 30 FC by 100W deluxe bright white mercury vapor lamp downlights. This visually separates the busy activity of the main corridor from the more quiet environment of the library.

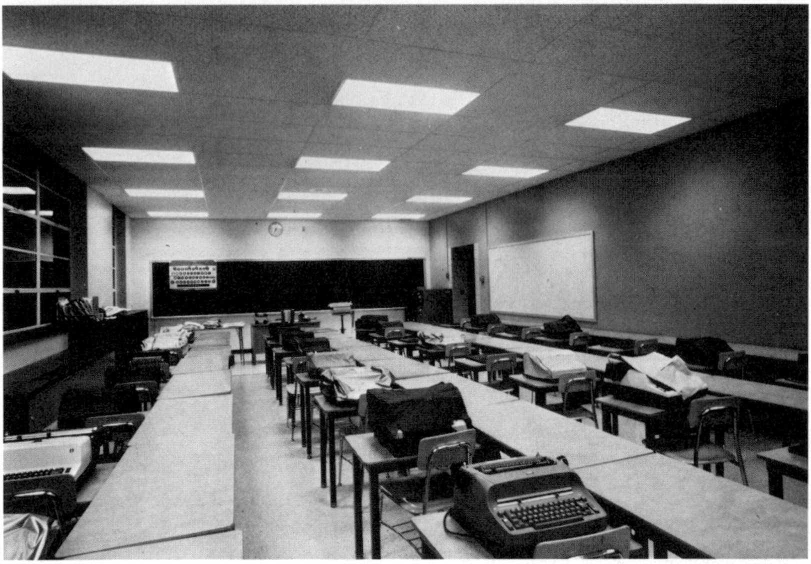

Figure 9-47. This is a typical typing classroom that is illuminated to 100 FC's in the high level 4-lamp mode.

Figure 9-48. This same room utilizing only the two inboard lamps is producing 53 FC's.

fixture can be turned off, reducing lighting levels and power consumption by 50%. (See Figure 9-48.)

The overall lighting design for Richard Montgomery High School not only took into account high visual comfort probability but incorporated the prime requirement considerations of task illumination, lighting effectiveness factor and contrast rendition factor. The cutoff of the light distribution resulted in high VCPs. During design, consideration was given to the following factors: visibility by proper intensity of lighting, comfort through control of brightness and glare, depth by allowing various light directions to give a third dimension to the space, verisimilitude, contrast for warmth, concentration and diffuseness, and quality of objects lighted. The overall electrical installation cost was only 13.64% of the total rehabilitation; the system is also proving low in operating cost.

ADDITION AND RENOVATION: SENECA VALLEY HIGH SCHOOL, MONTGOMERY COUNTY, MARYLAND

Seneca Valley is similar to Richard Montgomery in many ways. The 33,400 square foot addition and 11,140 square foot renovation cost $1.35 million. The school also features a subdividable master gym, but also includes three art studios. The design of this facility also complies with the current BOCA Energy Conservation Code and ASHRAE 90-75.

In an area of 5,700 square feet with a ceiling height of 19 feet, the gymnasium has a maintained 72 footcandles of illumination by the use of 24 continuous, full-range dimming, 400-watt SMH lighting fixtures offering 100 lumens per watt and 20,000 hours of life. (See Figure 9-49.) The even illumination is attested to by the light scalloping on the walls at a height of approximately 10 feet. (See Figure 9-50.) The energy use in this gym works out to 1.97 watts per square foot or 36 fc per watt. This gymnasium can be partitioned into two separate gyms, each with its own independent lighting and sound system. There are 68 footcandles when divided and 72 footcandles when opened into a joint-use room. In the divided mode, we obtain 34 fc per watt as opposed to the open mode at 36 fc produced per watt.

Again, fluorescent lighting is used with the MH lamps for good color quality and for emergency lighting, so that an existing emergency

Figure 9–49. The new main gymnasium can be partitioned into two separate gyms. Each has its own independent lighting and sound system. In this view there are 72 FC's.

Figure 9–50. With the AEC system set for a lower illumination level, we now have only 30 FC's for a different function use.

generator rather than a new UPS or inverter/rectifier system could be used.

Since fluorescent classroom lighting was used, optimum energy conservation and lamp life were realized through the use of a 20,000 Hertz fluorescent dimming system in the art studios. (See Figure 9–51.) Like the AEC system used for HID lighting, the 20,000 Hz dimming system saves energy in direct proportion to lighting level reductions. 50% of maximum possible illumination consumes 50% energy. In addition, by operating fixtures at 20,000 Hz rather than the conventional 60 Hz, this dimming system extends lamp life (thus reducing maintenance costs) by 20%. The 20,000 Hz system also uses only 80% of the electrical energy of a conventional non-dimmed fluorescent system in order to produce the same lighting level of illumination, whether or not the fixtures are dimmed. (See Figure 9–52.) This system has the capability, similar to the HID AEC system, to reduce lighting levels in direct proportion to available natural light. In the art studios totalling 3,350 ft^2, the additional first cost of the 20,000 Hz system was $7,414 in order to control 55 2′ × 4′ 4-lamp fixtures.

Figure 9–51. This is one of several art studios that utilizes a 20,000 Hertz dimming and automatic energy control system in conjunction with a photocell for feedback zone control of the lighting fixtures.

Figure 9–52. The illumination level can be dropped in the Art Studio. When it is used for nighttime adult classes, the AEC system calls for more electric lighting from the overhead fluorescent lighting system since there is no daylight contribution to defray its utilization.

Figure 9–53. This computer classroom utilizes computer relay terminals (which are cathode ray tubes) and word processor units. It incorporates a Hi-lo lighting system by switching inboard and outboard fluorescent lamps and ballasts separately.

Yearly energy savings for both lighting and air conditioning were projected at $861 for a payback period of 8½ years.

Classrooms in general throughout Seneca Valley employed alternate ballast switching to control fluorescent lighting. (See Figure 9-53.)

At Seneca Valley, as in the foregoing projects, all fluorescent and HID fixture ballasts were, as a matter of course, specified with a rated power factor of over 90%, to maintain high building power factors.

CONCLUSION

Each of these buildings offered an opportunity to develop innovative lighting solutions—solutions in harmony with the building design, the owner's considerations of corporate image, and energy conservation. In each project, lighting was integrated with the structure in a decorative and energy-conservative, yet effective manner. Levels of illumination in all cases conformed to or surpassed the recommendations of the then current IES Lighting Handbook.

Specific techniques and concepts used included service to lighting fixtures at 277 volts where possible. The electrical design of some of these projects resulted in custom lighting without custom construction. All components were items of proven dependability, low initial cost, low maintenance cost and low energy consumption.

All of these projects have illustrated the extent to which time-proven, off-the-shelf items can be employed by the modern electrical design engineer to develop energy conservative interior lighting systems, and lighting systems with low overall life-cycle costs. The simple provision of adequate lighting at relatively low levels of energy consumption is no longer a technical problem as it was only a few years back. Before the energy crisis became a mundane reality, lighting was routinely designed to levels 50% higher than today's, and they were energy inefficient with perhaps 200% more energy consumption. The challenge facing today's design engineer is rather in the quality and control of lighting, and at further reductions in power consumption. Particular techniques emphasized included the use of highly efficient lamp types, of air handling fluorescent fixtures (heat-of-light concept), and control of lighting with automatic dimming. The intent of all of these techniques is the reduction of energy use by, and life-cycle cost of, interior lighting systems.

10 ENERGY CONSERVATION IN A COMPUTER FACILITY AND A MAJOR PRODUCTION KITCHEN'S LIGHTING SYSTEMS BY AUTOMATIC LOAD-SHEDDING TECHNIQUES

FAIRLAND DATA CENTER, BUILDING NO. 2, A C&P TELEPHONE COMPANY COMPUTER FACILITY

Within the last few years, the tools and technology available to the lighting designer have increased. New lamps, lenses and highly effective fixture designs, in conjunction with flexible controls such as automatic dimming and sound-actuated switching systems offer the informed engineer unparalleled opportunities to provide optimum lighting at minimum levels of energy consumption and at reduced life-cycle costs, compared with older lighting installations.

However, new techniques incorrectly applied may result in poorer performance at increased cost, sacrificing efficiency as well as energy conservation and improved operation for the sake of technical advancement. This tendency has been observed in many professional disciplines.

In contrast, care and consideration in tailoring available techniques and technologies to actual needs can result in efficient, energy-conservative and relatively inexpensive solutions to highly complex problems. Past experience, more often than not, shows us that many times the elegant solutions are often the simplest.

This is as true in electrical design as in other fields. One example is the lighting system for Fairland Data Center, Building No. 2. This facility is one of many Chesapeake and Potomac Telephone Company computer facilities. Now in service nearly seven years, the electrical design is basically conventional and uses none of the lamps and control-system technology that have become available recently. Yet, it is uniquely tailored to its purposes, and is closely integrated with the overall electrical system design and installation. It was designed for flexibility and dependability—the design criteria for all the electrical systems, including lighting.

General lighting is provided by 2′ × 4′, 4-lamp, 40-watt, rapid-start, cool white fluorescent lamp grid troffers that, due to the lens selection, offer high visual comfort probability and high equivalent sphere illumination. The illumination is 150 field-measured footcandles, with over 100 ESI footcandles, while consuming slightly less than 3.8 watts/ft^2 or 39.5 fc/watt.

AUTOMATIC LOAD SHEDDING

Central to the lighting design are separate branch feeds to each of the two ballasts in the four-lamp luminaires. This separate ballast feed has broad implications in relation to the design of the overall electrical distribution system

The heart of the energy controls for this complex is a system of plug-in, bus-duct power taps that utilize both normal utility power and turbine generator emergency power. A double-section panelboard on each floor is fed from the normal and emergency power feed bus-duct. A contactor switch is installed within this split-bus panelboard at midsection. All lighting branch circuits for the floor's computer section and offices are fed from this panel.

Figure 10–1 schematically illustrates this simple, yet effective

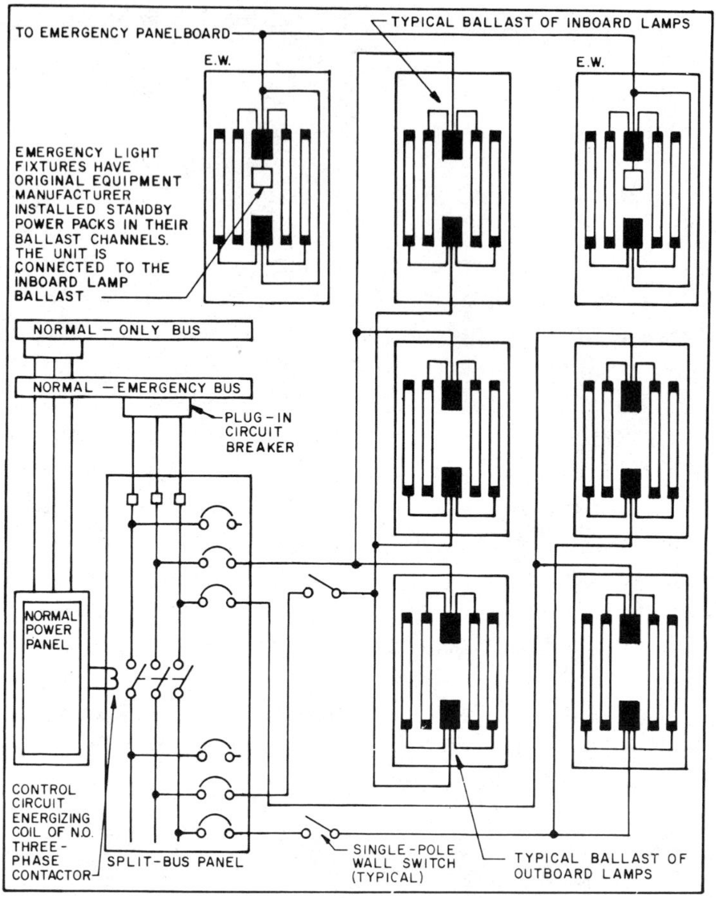

Figure 10–1. Schematic of load shedding system.

system. During a utility power failure, the lighting panelboard is fed turbine generator emergency power which is in limited capacity. This emergency generator power is principally reserved to power certain essential Class-A computers, uninterruptable power system (UPS) power, and Class-B computers. During a utility power failure, the contactor in the split-bus panelboard drops out half the branch circuits (See Figure 10–2.) With 50% of the lamps off, this not only reduces the lighting levels by half, but correspondingly reduces the power consumption by 50%. (See Figure 10–3.) Ballasts controlling inboard tubes are connected to a phase wire in the upper split-bus sec-

Figure 10–2. A double section panelboard on each floor is fed from a normal and emergency power feed bus duct, and feeds all lighting branch circuits for the floor's computer section.

Figure 10–3. During an emergency when only turbine power is available, illumination level drops to one-half by open circuiting the bottom half of the split bus panel.

tion, which receives normal utility power and which, during a utility power failure, receives emergency power. Ballasts controlling outboard tubes are fed from the lower split-bus section via a switch leg connected to a normal utility phase wire. This drops out automatically when the contactor senses utility power failure.

When a power loss is experienced, two 750 kVA turbine generators are energized to power the secondary switchgear. However, since the amount of emergency power is limited, the switchgear sheds most of its load by predetermined control device selection. This feature temporarily drops nonessential building systems out of service to reduce the total demand for electric power. When normal power is lost, it is necessary to sequentially phase emergency power back to the other systems critical to the building function—chillers, circulating pumps, cooling tower fans and air handling units—while sufficient power is available to meet those starting loads. The computers are picked up by turbine-generated power (over and above the UPS system) for the duration of the electrical failure. In this manner, the super-critical computers are not subjected to a power loss unless the turbines fail to come on the line within the battery reserve limits of the UPS system, which is normally 30 to 45 minutes.

For about three minutes, while the essential loads are being sequentially picked up during the transition from utility to emergency power, there is no power available for exit and emergency lights. Thus, certain select emergency fluorescent fixtures are actuated by the use of OEM integrally installed, standby power packages within their ballast channels. These units are a combination of rectifier, floating battery source, and inverter.

During an emergency, when only turbine power is available, the illumination level automatically drops by one-half, open-circuiting the bottom half of the split-bus panel. This technique is slightly more expensive than wiring alternate lighting fixtures on normal or emergency power, but does not produce spotty or uneven illumination during a power failure.

This technique also allows two normal levels of light as desired or required and allows energy conservation to be exercised at any time without sacrificing good, even illumination. (See Figure 10–4.) Should the lighting levels be deemed too high, they may be lowered by switching off that half of the lamps not backed up by turbine power. (See Figure 10–5.) Since these are not the same lamps that are supplied by

Figure 10–4. Computer area utilizing emergency lighting and power shedding hi-low automatic switching system at highest illumination level.

Figure 10–5. Same computer space as above in Figure 10–4 but illuminated at 53% of previous setting.

utility/emergency power, an area will never be left in total darkness. To shut an area down entirely, the lighting fixtures may be switched off at the panelboard by their respective branch-feeder circuit breakers. If energy conservation is not to be left up to individual choice, then the bottom half of the panelboard can be open-circuited by deactivating power to the panel's mid-level contactor. This will simulate a power failure and drop out half the branch circuits, thus extinguishing half the lamps in the fixtures (the outboard lamps). By removing the outboard lamps from operation, a reduction in lighting level of approximately 47% will be brought about, with a corresponding reduction in energy use of 50%.

During normal power operation, the computer technician has the option of having all the lamps on to give a high, even illumination of 150 footcandles. (See Figures 10–4 and 10–6.) If that level of illumination is not required, he can switch off half the tubes in each fixture, reducing the lighting level to 80 fc. (See Figures 10–3 and 10–5.)

Figure 10–6. At Fairland Data Center Building No. 2 in Washington, DC, 2′ by 4′, 4/40-watt rapid start cool white fluorescent lamp grid troffers provide soft light, and high VCP and ESI.

CONVENIENT SWITCHING, ENERGY CONSERVATION AND AUTOMATIC LOAD SHEDDING

Some executive personnel are located in future computer spaces designed with the same lighting and energy conservation techniques. Because of the office partitioning, lighting levels are reduced to 80 footcandles in the four-lamp mode (see Figure 10–7) and 40 footcandles in the two-lamp mode. (See Figure 10–8.) Another feature of this lighting design, in addition to separate branch feeds, is the ability to switch the twin ballasts in each fixture separately.

The overall lighting design for this project not only took into account high visual comfort probability, but the consideration of task illumination, lighting effectiveness factor and contrast rendition factor. The fixtures selected were chosen for their low initial and low operating costs. Lighting in this installation was in accordance with then current IES Handbook standards for illumination and brightness contrast. In other areas, design levels were higher than IES standards, but were obtained at lower energy consumption levels than if a conventional design had been used at lower levels of illumination.

Figure 10–7. Executive personnel office space located in an area designated for future computer use illuminated to full 100% level.

Figure 10–8. Executive personnel office space located in an area designated for future computer use utilizing 53% of previous illumination level.

The design was value-engineered based upon initial cost, cost of capital (principal and interest), maintenance, availability of the equipment used, and operating costs. The total overall electrical design for this computer facility was low in initial cost (13% of the overall cost) and is proving low in operating cost. This is due, in part, to the use of conventional off-the-shelf equipment and to manual and automatic energy conservative techniques.

Design According to the Code

The National Electrical Code, Article 384–16(c) requires that "the total load on any overcurrent device located in a panelboard shall not exceed 80 percent of its rating where in normal operation the load will continue for 3 hours or more," except "where the assembly including the overcurrent device is approved for continuous duty at 100 percent of its rating."

Thus, a typical 20-ampere circuit breaker is rated for 16 amperes at continuous duty. At 265 volts then, it can carry 4240 watts of energy (265 volts × 16 amperes = 4240). Each 4-lamp, 40-watt, fluorescent fixture has two 2-lamp ballasts utilizing 92 watts each, or 184 watts per fixture. The NEC requires at least one neutral for every group of 3-phase wires, which will reduce the neutral current to zero for a purely resistive, 3-phase, balanced-load, wye-connected circuit, but in a 3-phase, wye-connected fluorescent circuit, the neutral current will be nearly equal to the phase currents because of ballast third harmonics.

In this split-bus panelboard design, six phase wires share a single neutral, but there is no code violation. There are two conductors per phase and each carries only 2024 watts so that each pair of phase conductors is carrying 4048 watts, well within the continuous rating of the 20 ampere #12 AWG neutral conductor and of the 20-ampere circuit breakers. One of each pair of phase wires is connected to the top section of the panelboard's normal/emergency power and is wired to 22 ballasts of the inboard lamps; the other phase wire is connected to the bottom section of the panelboard's normal utility power and is wired to 22 ballasts of the outboard lamps.

Another facility using a similar design technique is the Pentagon production kitchen and back-up administrative office space. An automatic lighting controller and energy shedder of greater sophistication was used in conjunction with a time clock. (See Figure 10–9.) This hybrid technique automatically controls the lighting levels and sheds unnecessary and unwanted lighting and energy consumption during periods of low-level activity. There are many control systems on the market for different lighting needs that can save energy, operating and maintenance costs. Many of these systems are ultrasophisticated, nonstandard and high in first cost. However, some low-cost, standardized, off-the-shelf item was needed in order to increase lamp life, conserve energy and make effective use of the work-cycle concept of lighting needs in conjunction with the general area lighting system utilized. (See Figure 10–10.) Where possible, it should use an automatic load-shedding technique or system designed to enhance efficiency and extend the life of lamps, ballasts and other lighting fixture components. The following analysis is about just such a system.

TIME-CLOCK CONTROLLED AUTOMATIC LOAD SHEDDING

Individual ballast control can be used to provide a simple yet effective method of automatic lighting and load shedding capability. Each kitchen lighting panelboard is of the split-bus type and has two normally

Figure 10–9. Split-bus, split section, contactor controlled panelboard.

open contactors, one on the incoming section of the panelboard and the other at its midsection. (See Figure 10–9.) Both contactors are controlled by a time clock. (See Figures 10–9 and 10–10.) Ballasts controlling inboard tubes are fed from a circuit breaker in the top half or incoming section of the bus of the panel and ballasts controlling outboard tubes are fed from the lower or bottom half of the split-bus section via a switch leg. (See Figures 10–11 and 10–12.)

The bottom half of the panel and thus the outboard half of the

Figure 10–10. Split-bus, time clock activated double section panelboard.

lamps extinguish automatically when the midsection contactor has its power cut off via the time-clock control contactor. The inboard lamps are extinguished by the dropping out of the incoming section contractor being open-circuited when the time clock cuts off control power to it.

Time-clock controls can activate and deactivate lighting at predetermined times that allow for a higher degree of reliability. Time clocks are also provided with 16-hour reserve power which can carry

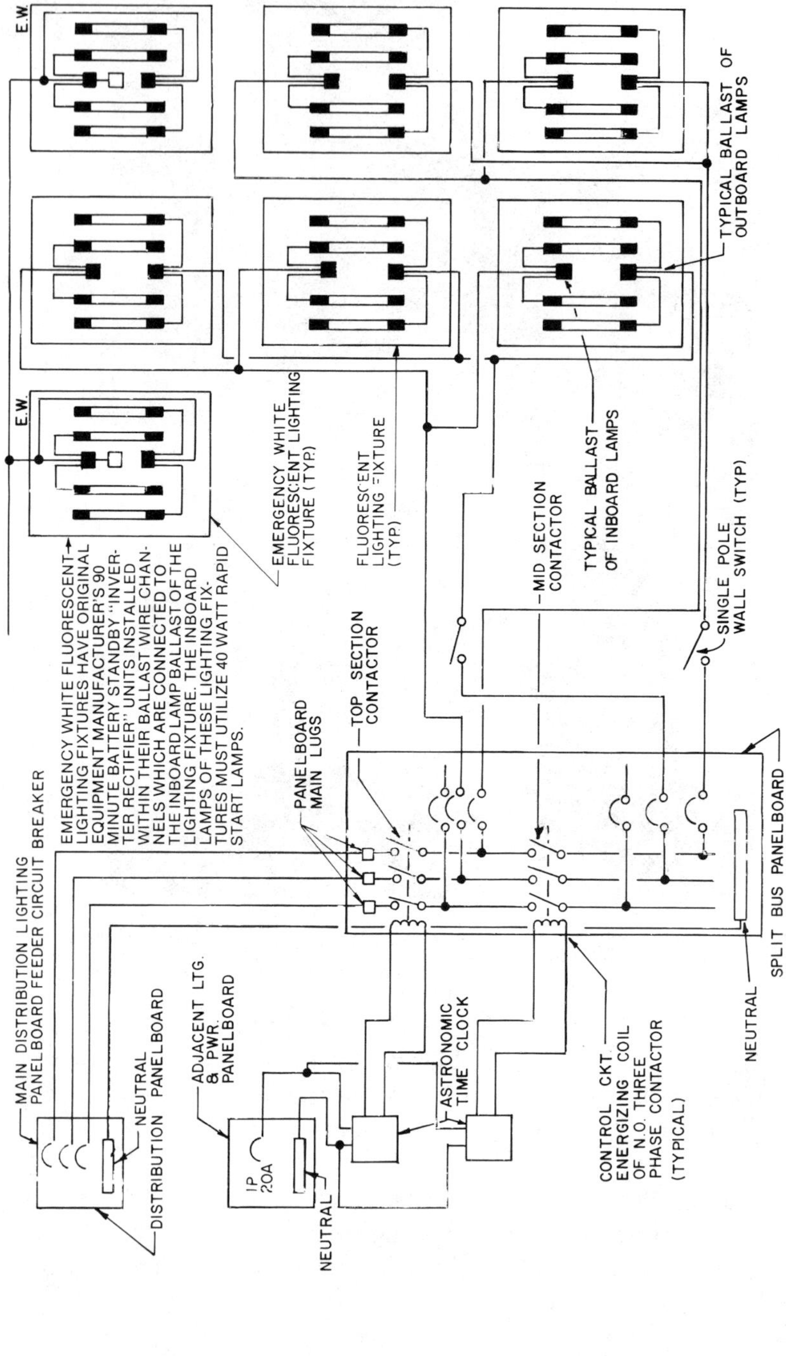

Figure 10-11. Lighting emergency load shedding detail.

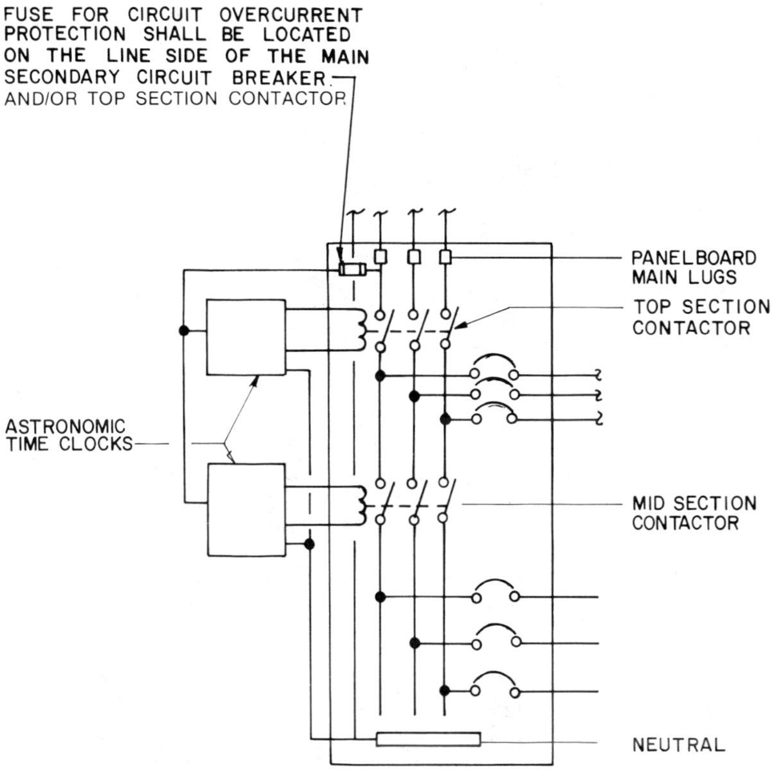

Figure 10–12. Alternate time clock feed detail.

the clock automatically through a power outage or brownout (reduction in voltage) and thus prevent loss of accuracy. A day-omitting feature is also available to prevent automatic operation of lights on weekends and holidays.

The schedule for the Pentagon has the following work shifts: the bakery hours are from 12:00 midnight until 1:30 P.M. (see Figure 10–13); kitchen hours are from 5:30 A.M. until 2:30 P.M. (see Figure 10–14); the potwash areas's hours are from 6:00 A.M. until 4:30 P.M.; and the office hours are from 8:00 A.M. until 5:00 P.M. (see Figure 10–15). In the bakery, kitchen and potwash areas the cleaning period is from shutdown time for that particular area until 10:00 P.M.

As a result, the work and clean-up lighting time schedule was based upon the following: full intensity (100% illumination) comes on in an area one hour before the start of that work shift and remains on until

Figure 10–13. Bakery area.

Figure 10–14. Steamer cooking area.

Figure 10–15. Production manager's office.

one hour after that work shift. Half the illumination levels are used for clean-up crews from one hour after the area work shift shuts down until 10:00 P.M. Then all the lamps and general illumination goes off until the next work shift, except for the emergency white standby lighting and exit lighting. (See Figure 10–16.) All the fluorescent lamps come on for the bakery at 11:00 P.M. until 2:30 P.M.(see Figure 10–17), and then the lower panelboard contactor drops out, thereby extinguishing the outer two lamps so that approximately 53% of the full illumination level is left on for the clean-up crews from 2:30 P.M. until 10:00 P.M. (See Figure 10–18.) All lamps are on for 100% full lighting output in the kitchen from 4:30 A.M. until 3:30 P.M. (see Figure 10–19), and then 53% of the illumination level is left on for the clean-up crews from 3:30 P.M. until 10:00 P.M. The potwash area has full 100% illumination from 5:00 A.M. until 5:30 P.M. and then shifts to 53% illumination from 5:30 P.M. until 10:00 P.M.

If the lighting panelboard is rated at 400 amps main lugs only (MLO) by being subfed from a 400-amp main circuit breaker, then its main incoming contactor switch must be rated for 400 amps since it

Figure 10–16. Emergency white and exit light egress lighting.

must carry the full load of the panel (both upper and lower bus capacities). The midsection contactor is then rated for just half the load or 200 amps since it is controlling and carrying only half the current in amperes. When the time clock calls for all the lamps to come on in all the fluorescent lighting fixtures, then both the incoming and midsection contactors close simultaneously. At this time the midsection contactor will be carrying half the current (200 amps) and the incoming section contactor will be carrying the full panel current of 400 amps. One hour after the end of the work shift the midsection contactor open-circuits, extinguishing the outboard fluorescent tubes. At the

Figure 10–17. Mixing machines.

Figure 10–18. Mixing machines.

Figure 10–19. Vegetable preparation.

end of the clean-up period the incoming section contactor will drop out upon command from the time clock. At this time the incoming section contactor is carrying only half the full-load current in amperes when it open-circuits.

If the panelboard is rated for 300 amps (a 300 amp MLO panel), then the incoming section contactor must be rated for 300 amps in order to carry the full rated current. The midsection contactor then must be rated for 150 amperes of current since it would never have to carry more than half the rated load current of 150 amperes. In actuality, the 400-, 300-, 200- or 150-amp contactors are not intended to carry those exact corresponding load currents but are oversized 125% according to code in order to carry a continuous 80% load capacity.

Article 384–16(c) of the National Electrical Code requires that "the total load on any overcurrent device located in a panelboard shall not exceed 80 percent of its rating where in normal operation the load will continue for 3 hours or more," except "where the assembly including the overcurrent device is approved for continuous duty at 100 percent of its rating." Thus, the 400-amp rated contactor must be capable of carrying 320 full load amps continuously and the 300-amp rated contactor must be able to carry only 240 amps continuously, etc.

As can be seen from Figure 10–20, a very substantial energy savings will be accrued through use of the Automated Hi-Lo time-clock actuated switching system. Depending on the area, number of lighting fixtures and the hours of operation at reduced energy consumption, a payback of approximately three years can be expected for the additional wiring, flexible modular wiring controls, switches, time clocks and contactors.

Even though energy-saving lamps and ballasts are used, which helped to minimize the initial load, the additional energy and cost savings accured with the automated Hi-Lo system are phenomenal. Although lighting related energy savings are large, and at no additional initial investment (Figure 10–21) the energy-saving ballasts showed only minor savings (See Figure 10–22.) This list price difference between a 40-watt, RS lamp at 3150 lumens and a 34-watt, RS lamp at 2850 lumens is 45¢. However, when they were bought in bulk there was no difference in the wholesale price they were purchased at. For this facility the energy-saving ballasts will require approximately a six-year payback period. (See Figure 10–22.)

SPECIFIC DUAL-LITE WIRING METHOD DETAILS OF OPERATION

1. The 2′ × 4′; 4-tube fluorescent fixtures are switched in the following manner (see Figure 10–23):

   - The ballast of the inboard tubes is continuously fed from a phase wire connected to the top half of the split-bus panelboard and receives normal utility power through the top contactor.
   - The ballast of the outboard tubes is fed by a switched switch leg fed by a normal utility phase wire connected to the bottom half of the split-bus panelboard and receives normal utility power through the centrally mounted (bottom) contactor.

AUTOMATED HI-LOW LIGHTING FIXTURE SWITCHING

AREA	NUMBER OF LIGHTING FIXTURE AT 150 WATTS EACH	** TOTAL HOURS OF REDUCED LIGHTING LEVELS AND ENERGY CONSUMPTION PER YEAR	ENERGY SAVING PER LIGHTING FIXTURE IN WATTS	TOTAL ENERGY SAVING FOR THE AREA PER YEAR IN KW	COST SAVINGS IN ENERGY PER YEAR ON 6¢/KHW
BAKERY	108	1950	75	8.16	$ 954.72
KITCHEN	50	1690	75	3.75	$ 380.00
POT WASH	44	1170	75	13.35	$ 231.66
TOTALS	202	4,810		25.26	$1566.38

** HOURS BASED ON 260 DAY OF OPERATION PER YEAR

NOTE: THE INITIAL FIRST COST OF IMPLEMENTING THIS MEASURE IS AN ADDITIONAL $5,000.00.

Figure 10–20. Pentagon value engineering study.

ANNUAL ENERGY COST SAVING USING ENERGY EFFICIENT LAMPS							
AREA	SCHEDULED OPERATION TIME SPAN	HOURS OF OPERATION PER DAY	NUMBER OF LAMPS	** HOURS OF OPERATION PER YEAR	ADDITIONAL LAMP COST	ENERGY SAVINGS PER YEAR IN KWH'S BASED ON WATTS/ LAMP	COST SAVINGS IN ENERGY PER YEAR BASED ON 6¢/KWH
BAKERY	11:30 PM-2:30 PM	15.0	432	3900	$0.00	10,108.80	$ 606.53
	2:30 PM-10:00 PM	7.5	216	1950	$0.00	2,527.20	$ 151.63
	E.W. LIGHTS	24.0	20	8760	$0.00	1,051.20	$ 63.07
KITCHEN	4:30 AM-3:30 PM	11.0	206	2860	$0.00	3,534.96	$ 212.10
	3:30 PM-10:00 PM	6.5	106	1690	$0.00	1,074.84	$ 69.49
	E.W. LIGHTS	24.0	8	8760	$0.00	420.48	$ 25.23
POT WASH	5:00 AM-5:30 PM	12.5	192	3250	$0.00	3,744.00	$ 224.64
	5:30 PM-10:00	4.5	104	1170	$0.00	730.08	$ 43.80
	E.W LIGHTS	24.0	2	8760	$0.00	105.12	$ 6.31
TOTALS			1,283	41,100	$0.00	23,296.58	$1,402.80

** HOURS BASED UPON 260 DAYS OF OPERATION PER YEAR EXCEPT EMERGENCY WHITE (E.W.) LIGHTS WHICH OPERATE 365 DAYS PER YEAR, 24 HOURS PER DAY.

Figure 10–21. Pentagon value engineering study.

ANNUAL ENERGY COST SAVING USING ENERGY EFFICIENT BALLASTS

AREA	SCHEDULED OPERATION TIME SPAN	HOURS OF OPERATION PER DAY	NUMBER OF BALLASTS	** HOURS OF OPERATION PER YEAR	ADDITIONAL BALLAST COST	ENERGY SAVINGS PER YEAR IN KWH'S BASED ON WATTS/ BALLAST	COST SAVINGS IN ENERGY PER YEAR BASED ON 6¢/KWH
BAKERY	11:30 PM-2:30 PM	15.0	216	3900	$ 756.00	3369.60	$202.18
	2:30 PM-10:00 PM	7.5	108	1950	$ 378.00	842.40	$ 50.54
	E.W. LIGHTS	24.0	10	8760	$ 35.00	350.40	$ 21.02
KITCHEN	4:30 AM-3:30 PM	11.0	103	2860	$ 360.50	1178.32	$ 70.70
	3:30 PM-10:00 PM	6.5	53	1690	$ 185.50	358.28	$ 21.50
	E.W. LIGHTS	24.0	4	8760	$ 14.00	140.16	$ 8.41
POT WASH	5:00 AM-5:30 PM	12.5	96	3250	$ 336.00	1248.00	$ 74.88
	5:30 PM-10:00 PM	4.5	52	1170	$ 182.00	243.36	$ 14.60
	E.W. LIGHTS	24.0	1	8760	$ 3.50	35.04	$ 2.10
TOTALS			643	41100	$2250.50	7765.56	$465.93

**HOURS BASED UPON 260 DAYS OF OPERATION PER YEAR EXCEPT EMERGENCY WHITE (E.W.) LIGHTS WHICH OPERATE 365 DAYS PER YEAR, 24 HOURS PER DAY.

Figure 10–22. Pentagon value engineering study.

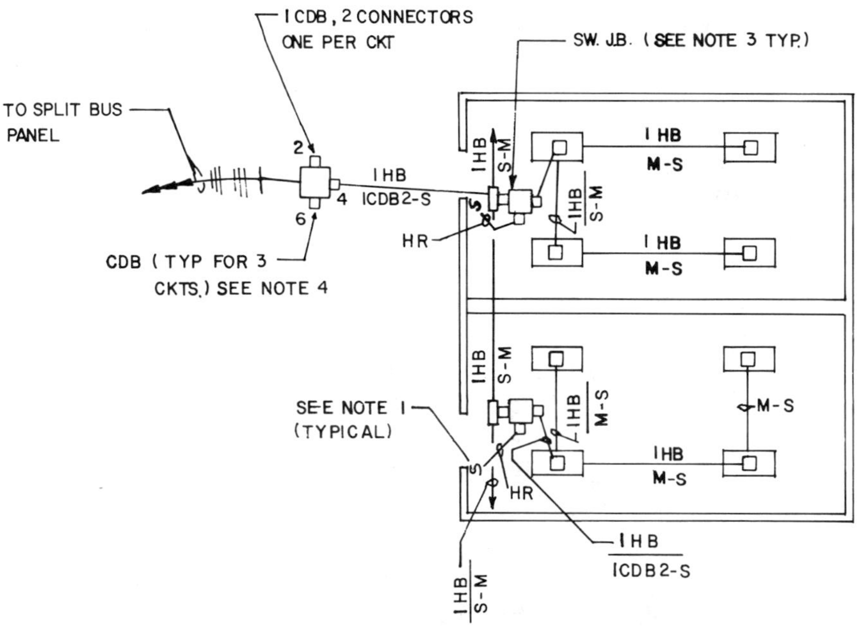

Figure 10–23. Dual-lite wiring method.

2. The 2′ × 2′ fluorescent fixtures utilizing two U-shaped tubes are switched with the inboard tube ballast of the 2′ × 4′, 4-tube fixtures.
3. The switch junction box is factory prewired with one each #1F4, 1CDB and #MS1 connectors which are all uniquely junction-keyed to prevent installation error. One is required for each local switch.
4. The contractor must connect the upper-panel part of the circuit to the black lead of the 1CDB2 connector for the inboard lamp activation and operation, and the lower panel lead (same circuit) to the red lead of the 1CDB2 for the outboard lamp activation and operation.

Each conventional 2-lamp, 40-watt ballast in conjunction with two 40-watt lamps consumes 92 voltamps (watts at unity power factor) of electrical energy. The same ballast with two 34-watt, energy-saving lamps consumes 80 watts. A 2-lamp, energy-saving ballast in conjunction with two 34-watt, energy-saving lamps consumes approximately 75 watts. The lighting circuit loads are calculated as follows: (1) each 2′ × 4′, 4-lamp, 40-watt, conventional 2-lamp ballast and 2 40-watt lamp combination consumes 182 watts; (2) each 2′ × 4′, 4-lamp, 34-watt, conventional 2-lamp ballast and energy-savings lamps consumes 160 watts; and (3) each 2′ × 4′, 4-lamp, 34-watt, energy-saving 2-lamp ballast and 34-watt, energy-saving lamp consumes approximately 150 watts.

The National Electrical Code Article 384–16(c) requires that "the total load on any overcurrent device located in a panelboard shall not exceed 80 percent of its rating where in normal operation the load will continue for 3 hours or more," except "where the assembly including the overcurrent device is approved for continuous duty at 100 percent of its rating." Thus, a typical 20-amp circuit breaker is rated for 16 amps for continuous duty. A single-pole, 20-amp, 120-volt, branch circuit breaker can carry 1920 voltamps (watts at unity power factor) based on 120 volts times 20 amps times 0.8 demand factor as per the 1981 NEC. Thus, ten 2′ × 4′, 4-lamp, 40-watt, RS fixtures or twelve 2′ × 4′ fixtures with conventional ballasts and 4 34-watt RS energy-saving lamps and twelve 2′ × 4′ fixtures with energy-saving ballasts and 4 34-watt, RS, energy-saving lamps may be connected to one 120-volt, 20-amp branch circuit. Since lamps can be changed more readily than ballasts, the maximum number of 2′ × 4′, 4-lamp fixtures was limited to ten so as not to overload the circuit in the future, due to relamping.

The NEC requires at least one neutral for every group of three-phase wires which will reduce the neutral current to zero for a purely resistive three-phase, balanced-load, wye-connected circuit; but in a three-phase, wye-connected, fluorescent-system circuit, the neutral current will be nearly equal to the phase currents due to ballast third harmonics.

Although in this split-bus panelboard design six phase-wires share a single neutral, there is no code violation. There are two conductors per phase and each carries only 800 watts; so each pair of conductors is carrying 1600 watts, well within the continuous rating of the 20-ampere, #12 AWG, neutral conductor and of the 20-amp circuit breakers. One of each pair of phase wires is connected to the top contactor actuated section of the panelboard and is wired to 10 ballasts of the inboard lamps, and the other pair is wired to the bottom contactor actuated section of the panelboard and is wired to 10 ballasts of the outboard lamps.

CONCLUSION

This discussion of energy-conservative lighting control has considered alternate ballast switching, automatic load shedding and time-clock controlled lighting controls. The challenge facing electrical design engineers of lighting systems both now and in the future is effective energy-conservative control of these light systems. This subject takes on greater importance as the cost of energy increases. The two design techniques described in detail are cost-effective, efficient and energy-conservative.

This chapter has attempted to deal with the application of technologically advanced techniques as well as to cover some of the philosophical issues regarding what engineers can do for our society through the proper application of technology. We have chosen to stress energy conservation in conjunction with and in harmony with alternate energy fuels as described in the following chapter. In a very brief way it indicates alternate approaches to today's dwindling fuel supplies and suggests new vistas in dealing with the energy "shortage." Each of the preceding chapters has attempted to address energy conservation through good design practice without engendering great personal sacrifice on the part of the American people.

To a world going on increasingly shorter rations, these design techniques will allow us to have tomorrow's technology today, and at less cost than today's conventional design techniques. In this way, without undue sacrifice on the part of the individual, we can preserve today's way of life tomorrow.

11 ALTERNATE ENERGY FUELS: NOT JUST ENERGY CONSERVATION OR NEW ENERGY RHETORIC

INTRODUCTION

Energy conservation by such techniques as reducing lighting levels is not a valid solution nor a proper response. Diminished levels of illumination can and usually do reduce worker productivity, thereby increasing the hours required for a task. This in turn requires facilities to operate for longer periods of time and at ever greater expenses of energy. Productivity and nonreplenishable fuels are thus sacrificed on the altar of energy conservation. Despite the 1973 Arab oil embargo and the subsequent 1979–80 Iranian crisis, the federal government has yet to deal with the hard realities of energy conservation. Instead of formulating a comprehensive and effective energy policy, Congress, the administration and the bureaucracy have been fighting the energy "war" sporadically. Piecemeal legislation and regulations have proliferated, preventing serious energy policy consideration. This Band-aid approach can neither cure the malady nor prevent it from getting worse.

As an example, lighting of every sort accounts for only about 5 percent of the world's energy consumption. Even if the associated energy load for air-conditioning interior spaces (approximately 0.5 W of A.C. to remove each watt of interior lighting heat) is included, lighting-related energy consumes less than six or seven percent of the world's energy.

Not all buildings are air conditioned and not all lighting is for the interior. Exterior security, flood, parking lot and sports lighting do not have attendant air conditioning loads. President Johnson attempted to impress the American people with the earnestness of his concern for energy conservation by turning out the White House lights, an action which has been emulated ever since. It is probable that those most pleased with this reducing of facade and corridor lighting are muggers, rapists and purse snatchers.

Office managers are a great deal less impressed. When Social Security headquarters in Baltimore, Maryland, reduced its illumination levels from 100 fc to 50 fc, productivity fell off 28%. This not only increased the time and payroll required to maintain the same work load, but actually increased the energy bill. The building operating day was so much longer, once productivity per worker had been sacrificed on the altar of visible energy conservation, that the building's energy consumption nearly doubled.

The Social Security office, of course, was an older building, a relic from the days of cheap energy. With a government committed to energy conservation (new General Services Administration [GSA] standards specified energy uses of 55,000 Btu/ft^2 per year for new buildings and 85,000 Btu/ft^2 per year for rehabilitated structures), new buildings should be able to do better. Thus, an attempt was made to do so with the Norris Cotton Federal Office Building in Manchester, New Hampshire, completed October 8, 1976. It is an experimental building, with no two floors typical. Every sort of lighting, HVAC and energy conservation system known to modern technology was employed. Initially, it consumed more than 150,000 Btu/ft^2 per year. After a malfunctioning outside air damper was repaired, consumption was reduced to 100,000 Btu/ft^2 per year. Today, after six years of operation and further fine-tuning with an in-house computer system, it is approaching the GSA energy use standard for a rehabilitated building.

Meanwhile, then-energy czar Simmonds had a fireplace constructed in his Washington office space, allegedly to reduce use of the central heating system. This idea apparently was to heat spaces which he worked in and not all the unused adjoining spaces. Considering the resource utilization factors involved or natural resources consumed in processing, refining, transporting, converting and delivering the tradesmen and goods to make the fireplace, enough energy was probably consumed to require a 20-year payback on the cost of construction and a 40- or 50-year period of use to break even on the total life-cycle cost and energy consumed. The initial cost, expense of delivering wood to a downtown address and the notorious inefficiency of the average fireplace were apparently not considered either. To add insult to injury, several months later the energy czar was quoted as saying the energy crunch was over and he was having the fireplace removed, again at additional monetary and energy expense to the taxpayer.

All of these actions smack more of public relations rather than serious thinking about energy policy. They are microscopic rather than macroscopic approaches. To take a macroscopic approach, it is first necessary to take a broad look at where our energy comes from and how it is used.

The American energy use is supplied approximately as follows: 47% from oil; natural gas, 27%; coal, 18%; nuclear, 3%, and all other forms, 5%. Of the oil consumed, 55% goes for transportation, 18% for industry use, 17% for residential and commercial users, 9% for electric power generation and utilities, and 1% for all other uses.

A General Motors study breaks down average household consumption in this way: 34% for house heating, 32% for driving (15,000 miles per year), 8% for heating water, 7% for lighting and air conditioning, 5% for cooling or freezing foods, 2% for cooking; while clothes drying, color TV, small appliances, clothes washers and dishwashers each account for approximately 1%.

These breakdowns indicate the extent to which current conservation measures may be effective. Those aimed at transportation (55% of oil use) will obviously be the most effective. Those like the past 78°F thermostat settings for commercial establishments (with HVAC accounting for only a fraction of the 17% attributable to all commercial and residential uses) affect only a small portion of overall energy use and, thus, are obviously only marginally effective.

The simple fact is that energy conservation is not being handled wisely and comprehensively. (See Figure 11–1.) Conservation efforts should be augmented by a vigorous program aimed at creating and us-

Figure 11–1. Energy conservation is not enough in itself. Consumers can conserve by using less and using it wisely. But no matter how much we all conserve, it won't keep up with demand. By 1990, we may be able to save 20 percent with everyone's best conservation effort put forth, but at the same time, demand will have grown to 50 percent. Thus, although conservation will help, it cannot cover the total job of meeting demand. Even if increasing conservation techniques could meet the increasing demand based on a slow rate of population and economic growth (in order to maintain the standard of living and keep unemployment down), we would still be consuming the same amount of energy. In the long run, by continuing to utilize nonreplenishable fuel sources we will eventually run out, since there is only a finite quantity left. However, increased efforts to conserve energy could well buy mankind the time needed to develop and perfect second and third generation alternative replenishable energy sources and get them on the line in order to start production. (Photo courtesy of General Motors Futurama Exhibit.)

ing nondepletable energy resources. There are many approaches which may be taken:

Increase the use of nuclear fusion plants.

Get fast breeder reactors on the line quicker. (These are nuclear reactors which produce more refined fuel than they use.)

Increase the use of existing geothermal sources and speed research into the use of new sources such as geopressurized zones of molten granite (plutons).

Increase production of electrical energy through updating and replacing existing generators and hydroelectric turbines. (Hydro presently accounts for 10% of Washington state's power.)

Develop unconventional natural gas from geopressurized brine deposits.

Develop different types of wind machine technology.

Develop a technology for converting oil shale, coal, biomass, wood pulp and other types of energy into synthetic oil and gas. Wood waste can be converted to synthetic crude oil fuel in one hour instead of 275 to 345 million years. Synthetic oil's heating value of 15,000 Btu/lb is lower than the 19,000 Btu/lb of natural crude (due to the extra oxygen content of synthetic oil), but it could be converted to gasoline after the use of an additional step in refining.

Harness wave power. At least two methods, the Annapolis Experiment in induced draft wavepower generators and the use of flexible rafts that operate pumps to actuate hydraulic motors driving alternators are currently being investigated. (See Figure 11-2.)

Cultivation of oil-bearing and producing weeds such as milkweed, quackgrass, poinsettia, sow thistle and guayule. An oil bearing plant that produces high quality petroleum oil is the water hyacinth.

Production of hydrogen for fuel through the electrolysis of water. The molecular hydrogen H_2 and oxygen O_2 from electrolysis of water must be done without expending more energy to get it than the energy that can be extracted from the product fuel. Hydrogen and oxygen are nonpolluting fuels with water vapor and energy as their combusion byproducts. There are several modern photosynthetic processes in which photogalvanic electrodes are used to break water into its constituent parts. Another innovative procedure for electrolysis is the solar furnace designed by Howard Zackman, P.E.

Coal gassification to permit economical retrieval of deep (300 feet to 2000 feet down) coal. Gassification avoids the many problems associated with strip mining, such as pollution of surface and underground waters by acid waste created by water percolating through rock shale-waste back fill. Underground coal gassification was first developed by the Russians in the late 1920s.

The solar powered sterling engine. Parabolic reflectors are used to concentrate the sun's energy on the heat sink end of a sterling engine. The sun's energy is converted to heat energy, then to mechanical

Figure 11-2. As undersea living becomes truly feasible, the sea will yield its secrets to an army of determined explorers, revealing its unlimited promise as a source of food, chemicals, minerals and other natural resources. Recreational living in underwater hotels of the future will provide sportsmen the mobility of aquascooter transportation in an undersea world. Atomic-powered submarine trains will be seen plowing through the undersea world bringing cargoes of materials and supplies for the world's needs. Some of these sea trains will supply research facilities in such places as the hostile Antarctic environment. Special undersea drilling rigs will be working the oil rich deposits of the Continental Shelf. Oil that is piped from these oil rigs to undersea storage tanks will be transported to a nearby seaside refinery by submarine undersea oil train tankers. (Photo courtesy of General Motors Futurama Exhibit.)

energy and finally, through an engine generator connection, electrical energy is produced at 60 Hz output.

The magnetohydrodynamic (MHD) generator. High temperature (approximately 5000°F) exhaust plasma gases are produced by combustion of coal dust and air. This exhaust gas is seeded with metal ions of potassium or cesium to produce a highly saturated ion stream and freely disassociated valence electrons. As this highly charged ion gas stream passes by anodes and cathodes, a direct current is produced. The DC output can then be inverted to a staircase approximation of a sine wave and is then smoothed out by π or T filtering to produce a clean 60 Hz AC output.

Garbage to alcohol conversion, turning a liability and a costly disposal problem into an energy asset.

Conversion of grain crops, such as wheat or corn (the whole plant chopped into silage) and garbage into alcohol fuel sources (ethanol and methanol) for use in gasohol mixtures. (See Figure 11-3.)

Methanol can, however, corrode some materials now found in present day fuel systems. Gasohol, when exposed to water, has the deleterious effect of having the methanol separate out of the homogenized mix. Those processes of grain to fuel conversion do however improve

Figure 11-3. The arid desert blooms. Irrigated by desalinated sea water, the mineral rich desert sands nourish a myriad of crops and produce a number of harvests each year. Through such methods as intensive irrigation and hydroponics, the farms of the future will produce ever-larger quantities of food and fuel (through conversion of grains and agricultural waste processed into ethanol and methanol) for a growing world. (Photo courtesy of General Motors Futurama Exhibit.)

on nature's way of converting vegetable matter into (depletable) natural fuels such as coal and oil by a factor of 300 million years or so. (See Figure 11–4.) Currently, methanol can be produced for half the cost of ethanol and so appears more practical from a first-cost basis.

Conversion of methanol to synthetic gasoline, as in the Mobil Oil process, concentrates the hydrocarbon products (and potential energy) by wringing out the water component. By eliminating the 56 parts of water for the conversion to gasoline, the remaining 44 parts of hydrocarbon products leaves us with 95% of the Btu energy efficiency that was available in the methanol in the first place. This process is an improvement over using methanol itself in automobiles. Since methanol has only half the Btu content of gasoline, its sheer doubled bulk (to produce the same energy) would require production and distribution systems to be nearly doubled. However, there is a ray of hope hidden behind the clouds. Although methanol alcohol has only half the Btu content of gasoline, it generates 20% more energy through more efficient burning. A reciprocating internal combustion engine using methanol rather than gasoline burns cleaner by a factor of five to one and therefore longer engine life can be expected since alcohol burns cooler. This, in turn, reduces maintenance cost because there isn't any hydrocarbon build-up. An experiment two years back with modified

Figure 11–4. *WEATHER OR NOT.* By the 21st century man's ultimate control of his environment will be the predictable control of the planet's weather for the benefit of mankind. Mineral rich desert sands (once drought stricken) will, with the aid of weather control, nourish a myriad of crops and bring forth numerous harvests each year. (Photo reprinted with the permission of Manville Corporation.)

Ford carriers running on methanol produced 13 to 16 miles per gallon (depending on the driver), while its gasoline counterparts were getting 17 to 19 miles per gallon. The cost at the time for conversion to methanol fuels was between $600 and $800, and that cost is easily recaptured through substantially longer engine life. The conversion technology is simple enough—increase the compression ratio to approximately 12 to 1, redesign the fuel delivery system and change the timing. A larger fuel tank will also have to be installed, but it won't have to be doubled or treated with turnplate. In addition, gasketing materials have to be changed so as not to be adversely affected by the alcohol.

One additional benefit of burning alcohol is that the emissions from the tailpipe will be cleaner than the air we citizens breathe.

Solar power has yet to reveal its long-term promise since its developmental stage is still in its infancy. Costly local solar collectors for heating and hot water application are just the beginning. (See Figure 11-5.) The future will see the use of much larger and more efficient orbiting solar collectors to convert solar energy into electricity and beam it to earth in the form of microwaves. Each geostationary, geosynchronous collector would occupy a stationary parking orbit some 22,300 miles out in space beaming energy to fixed earth acquisition stations. In case of any orbital shift or abnormalities causing the "fixed" collector to shift its position, a fail-safe system would spread the normally narrow energy beam to prevent harm by reducing the energy density (energy input per unit area). Orbiting solar collectors would be both larger than earth-mounted collectors and more efficient due to their receiving solar energy directly, without the intervention of the atmosphere. The reason for several, preferably three, large solar collectors is that they would be more efficient units and would give complete coverage of the reclaimed energy to the entire globe. A comparative analysis is to consider the resource utilization factor or energy and raw materials that go into the production of millions of individual automobiles as opposed to manufacturing fewer busses capable of the same passenger carrying capacity. You end up with fewer pounds of mined and refined finished materials per person and more passenger miles per gallon. (See Figure 11-6.) The same applies to the use of many smaller, less efficient solar collectors on the roofs of residences and small offices, as opposed to fewer but larger solar collectors in earth orbit.

Figure 11–5. Energy from new sources will power the cities of tomorrow. Energy sources such as solar, geothermal and power beamed from geosynchronous geostationary solar collector stations orbiting high above the earth's atmosphere will be in use. Garbage disposal problems will ease as city waste is converted into ethanol and methanol for either gasohol or catalytic conversion to gasoline. The design concept as portrayed in this illustrated photograph is a self-reliant satellite industry fueled by the sun allowing an industrial complex to be sited in a natural forest preserve. A solar heat collector provides power and protects the area ecology while a monorail system hauls shipments of products. This is one of a series in "Innovation in Design," created for DAP Architectural Sealants which was designed and rendered by Harold R. Roe, A.I.A., Richard P. Howard Associates, Architectural Illustrators, Sylvania, Ohio.

Fortunately, RUFs which were discussed earlier and RIFs which are resource impact factors, are two relatively new cryptic energy acronyms that were dropped by the then new but soon to be deposed Department of Energy before everybody got completely burned up over

Figure 11-6. The making of tomorrow will include high speed, subsurface transportation to just about any place on the planet earth. It will help to bind giant cities together into megalopolises having greater identification and unification of its people and their values and traditions. (Photo reprinted with permission of Manville Corporation.)

Figure 11-7. On the frontiers of lighting technology. To a world going on increasingly shorter rations, the engineer brings tomorrow's technology today . . . to preserve today's way of life tomorrow. (Photo courtesy of General Motors Futurama Exhibit.)

them. They were untenable, unyielding and unworkable concepts and where RUFs were concerned, would have to be tabulated for every climatic zone and for every utility serving that zone, and would also have to be tabulated for every state, county and municipality having a different set of codes to design and build by.

CONCLUSION

These are the areas in which new technologies will be needed to fulfill the energy needs of the future, but to develop them to a point of economic feasibility will require quantities of willpower and money which have not yet been dedicated to them. To date the Government has appeared more fascinated with obscure energy bookkeeping techniques, with acronymic gadgetry. Resource utilization factors (RUFs) and resource impact factors (RIFs) yielded to building energy performance standards (BEPSs), which appeared no more workable but had the advantage of novelty; however, it too has passed into bureaucratic obscurity. Six months from now, another new abbreviation may hold the stage.

All of which just goes to show that then President Carter's characterization of our energy crisis efforts as the Moral Equivalent of War may be summed up in a single, very descriptive master acronym: **MEOW!** But MEOW is not enough; it is *action,* serious and concerned action which will be needed to develop the energy technology we need to preserve today's way of life tomorrow. (See Figure 11-7.)

12 A FUTURE OFFICE OF AN ENGINEERING AND ARCHITECTURAL FIRM

The following chapter provides a glimpse into an office of the immediate future utilizing some or all of the aforementioned energy saving and managing techniques. In addition it depicts not only the use of these concepts but their implementation into the design of other facilities by an engineering and architectural design firm.

The following is a possible scenario of a transaction in an engineering and architectural office of the not too distant future. Unlike the present day sign-in, check-in, reception function which is usually burdensome and antiquated at best, tomorrow's system will be sleek and functional. A visitor, consultant or prospective client, upon arriving at this modern and efficient office of tomorrow, will stop at the receptionist desk first and announce his or her name, business, purpose and the person with whom he/she has an appointment. The receptionist will then notify the host employee of his/her arrival while entering the visitor's supplied data on a computer relay terminal that has hard copy reproduction capabilities as well. Within seconds, the visitor will be issued a magnetic-card badge pass, that will graphically annunciate his or her location anywhere in the firm's premises within the building. This feature will allow the visitor free access of the facility without undue restrictions while also freeing the host escort or possible security guards of the responsibility of having to closely monitor the visitor's whereabouts. Should this individual venture too close to a security area, for which he has not been granted a previous clearance, the badge's magnetic trace will signal the building's security guards or other appropriate individuals or androids. (See Figure 12–1.) In addition, should this individual receive an emergency call during his/her visit, it will be possible to page that person at the exact loction within the building without disturbing other personnel and employees in the office complex. This same magnetic card will also be used as a monitoring system for employees in general, as well as providing access to the main entrance/exit system.

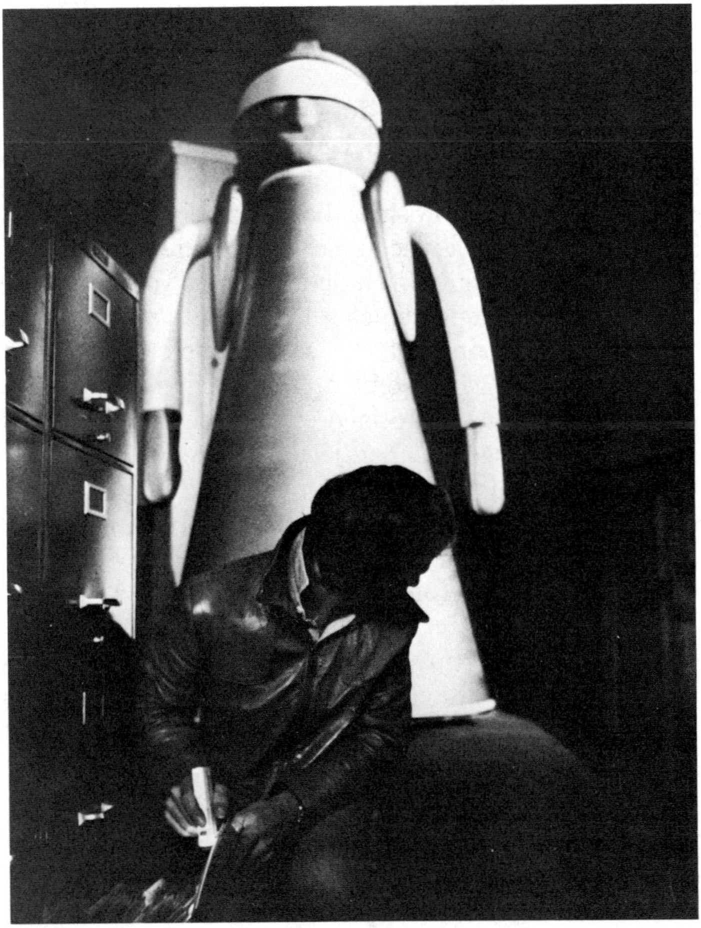

Figure 12–1. Sentry guard. (Courtesy Scantrak Corporation of Rutherford, New Jersey.)

Upon the receptionist notifying the guest that the person being visited is temporarily detained in conference, either an escort host or computer-programmed mannikin will come out and greet the guest, then take a beverage request and return with it. (See Figure 12–2.) This commercial android can then lead the guest to a luxurious and spacious multipurpose conference room, television studio, public relations showplace and interiors selection/demonstration workshop.

We are now in this future office and the android is explaining to the client in a clipped prerecorded humanoid voice that the partner in charge is temporarily delayed and will be with him shortly, and would the guest like to avail himself of the audiovisually entertaining public relations showplace?

"What is this place and how does it work?" he asks.

"Well, let me explain it to you," our host android chirps. "As you step before each of these individual observation portals they instantly light up and become windows into the future. When you step in front of each portal a photoelectric sensor above and below senses your presence by the interruption of a light beam as you pass through it. This initiates an automatic sequence that starts a prerecorded audio-visual presentation. Some of the scenes depicted are still, two-dimensional pictures, while some are dichromatic, polarized, twin-photo, three-dimensional pictorial presentations. In some cases a laser holographic three-dimensional projection is cast into the center of the room, and you can walk around and through the image and see what is behind every object or building. Each window portrays a different aspect of today's and tomorrow's design technology and will be indicative of the engineer of tomorrow being the architect and custodian of our technological future, building and bringing tomorrow's world today. Each window portrays design aspects that are on the frontiers of technology. To a world going on increasingly shorter rations, the engineer will bring tomorrow's technology today in order to preserve today's way of life tomorrow."

As undersea living becomes truly feasible, the sea will yield its secrets to an army of determined explorers, revealing its limitless promise as a source of food, chemicals, minerals and other natural resources. Recreational living in underwater hotels of the future will provide sportsmen the mobility of aquascooter transportation in an

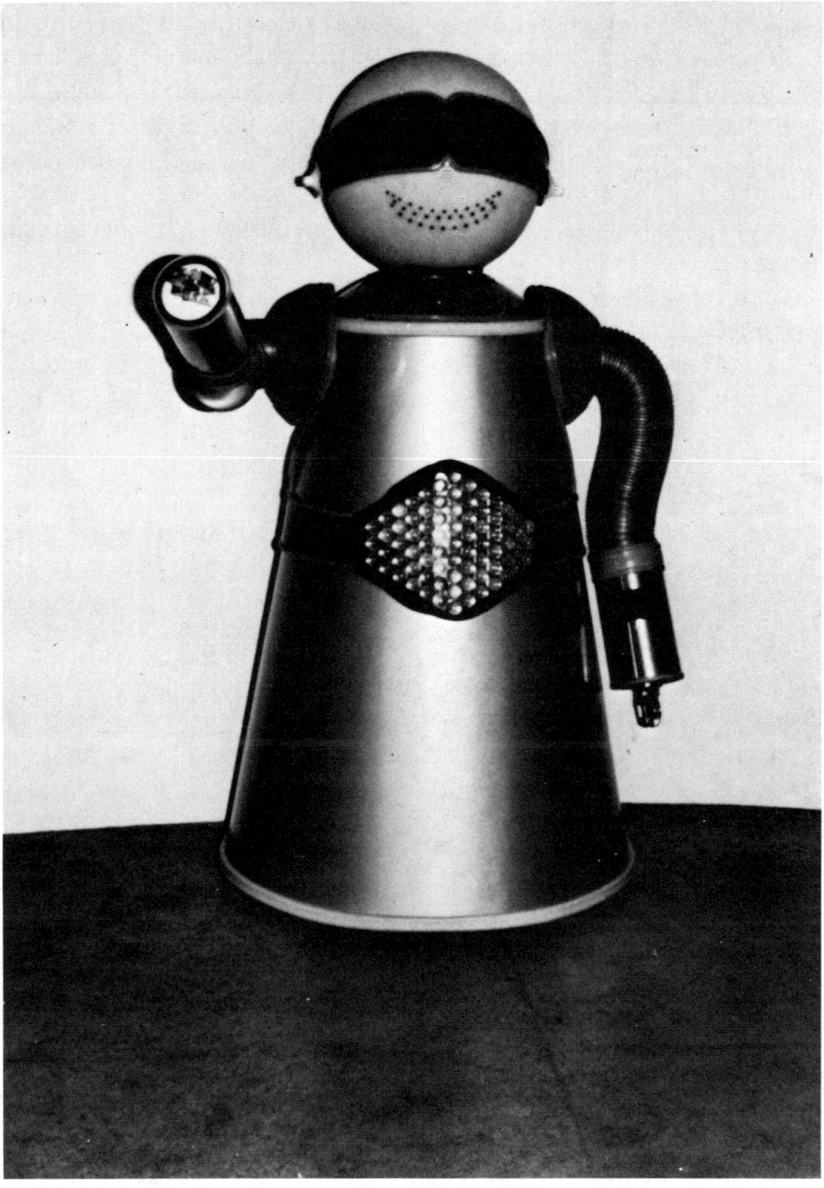

Figure 12–2. Commercial androids in the office environment of the future will function in various capacities such as a receptionist, escort host or security guard. (Courtesy Scantrak Corporation of Rutherford, New Jersey.)

undersea world. Atomic-powered submarine trains can plow through the undersea world bringing cargoes of materials and supplies for the world's needs. Some of these sea trains supply research facilities in such places as the hostile antarctic environment. Special undersea drilling rigs will be working the oil rich deposits of the Continental Shelf. Oil that is piped from these oil rigs to undersea storage tanks is transported by submarine undersea oil train tankers to nearby seaside refineries. (See Figure 12–3.)

Giant tree cutters, with laser beams for saws, clear the jungle for a superhighway of the future. They are followed by a mammoth self-propelled roadbuilder that leaves a finished elevated road in its wake. Two very special machines are in the process of felling 300-foot high tropical jungle forest trees. Another unique behemoth, equally

Figure 12–3. The engineer will be the architect and custodian of this world's future —building tomorrow's world today. (Photograph courtesy of General Motors Futurama Exhibit.)

strange looking and as tall as a building, follows behind these two electronic Paul Bunyans. (See Figure 12-4.)

A finished highway stretches in the wake of this jungle roadbuilder, a factory-on-wheels, which sets the foundation, forms the surface slabs, and lays them in one continuous operation. It is ingesting the toppled trees and other jungle growth and is leaving behind itself a paved multilane highway. Supply trucks and other vehicles are seen coming up the finished highway to supply building materials and equipment. This once wild, primitive jungle is converted into a useful, attractive jungle metropolis, dramatic proof of how mobility can help man master even the most primitive tropical environment. An area once attractive and useful to only the most adventurous has been turned into a productive contributor to the world's marketplace. (See Figure 12-5.)

Figure 12-4. Giant tree cutters, with laser beams for saws, clear the jungle for a superhighway of the future. They will be followed by a mammoth self-propelled roadbuilder that leaves a finished elevated road in its wake. (Photograph courtesy of General Motors Futurama Exhibit.)

Figure 12–5. A finished highway stretches in the wake of a jungle road-builder, a factory-on-wheels, which sets the foundation, forms the surface slabs, and lays them in one continuous motion operation. (Photograph courtesy of General Motors Futurama Exhibit.)

We are now in a remote mountain area where we are surprised to see a number of comfortable, livable homes. In one plateau area we are looking down into a valley between two sheer walls of adjacent mountain ranges. Perched on the sheer side of one of the mountains to the left of our view is a futuristic looking home. In the foreground is an outdoor swimming pool hacked out of solid rock in the face of the mountain. Stairs lead up to an enclosed indoor patio area leading to the circular home that is attached to the side of the mountain but nevertheless appears to be floating in space. Underneath the home and attached to it is an enclosed, glassed-in garage housing a small, sleek, energy-efficient car of tomorrow, electric-hybrid fuel system, fuel cell type—who knows? Leading out from beneath this garage of the future a roadbed serpentinely stretches down and around the mountainside and over to an enclosed elevated aerial roadway that bridges the gap between the two mountainsides. In the distance to one side of the mountain and in the distant valley we see many modern highways, allaying our first impression that the area is remote. We understand now how it is that working man can live wherever his fancies lead him, since he has so many fast arterial routes that keep him in touch with his work, his leisure and his fellow man. (See Figure 12–6.)

The arid desert blooms. Irrigated by desalted sea water, the mineral-rich desert sands nourish a myriad of crops and produce a number of harvests each year. Through such methods as intensive irrigation and hydroponics, the farms of the future will produce ever-larger quantities of food and fuel (through conversion of grains and agricultural waste processed into ethanol and methanol) for a growing world. (See Figure 12–7.)

All future farm corps are now practical to grow in the arid desert areas because man has learned to easily and economically convert sea water for irrigation purposes and readily pipe it to desert locations.

Figure 12–6. Today's engineers are in your future making today's ideas work for your tomorrows. (Photograph courtesy of General Motors Futurama Exhibit.)

Figure 12–7. Photograph courtesy of General Motors Futurama Exhibit.

Man's mobility and manipulation of his environment is evidenced by such technologically advanced devices as electronic cultivators, harvesters, planters, moisture sensors, seeders and other sophisticated devices that make desert farming efficient, practical and profitable. (See Figure 12–7.)

The city of tomorrow. A glittering complex of commerce and urban living, illustrating the use of air rights over existing thoroughfares. Low-rise dwelling units are located above garages, with commercial and recreational areas making up an inner-city community. One of the interesting things to catch your eye is that amid all this technologically advanced and futuristically styled architecture there nestles a simple and traditional Gothic church of the mid-twentieth century in the very heart of the city of tomorrow. (See Figure 12–8.) Yesterday's life styles and values preserved in the rapid mobility and changing lifestyles of tomorrow.

New concepts of terminal interchanges and the latest concepts and efficiencies in the handling of bus transportation make daily living a

Figure 12–8. Engineering that is dedicated to the events of today will influence man's social, economic, business and cultural lifestyles of tomorrow. Engineering designs and decisions that are on the "Frontiers of Information Technology" will bring purpose to progress. (Photograph courtesy of General Motors Futurama Exhibit.)

pleasant new experience. Highway freight uses the containerization concept now long in use with additional, more efficient concepts for better, swifter routing of all types of merchandise. Newly evolved freight terminals of the future will utilize new concepts of storage and rerouting of package containers as circumstances dictate. We noticed that as we approached the city of tomorrow on tomorrow's superhighway system, the thoroughfare acquired additional new lanes and picked up considerably more traffic. Despite all the traffic volume there is no confusion or delays, and the reason for this smooth flow of traffic is the use of an automatic highway system to facilitate man and free him of his labor. The cars move smoothly and freely as they drive on and off effortlessly on this road of tomorrow. Drivers and passengers can relax and read or talk with one another while their cars are in the automatic lanes, and just as assuredly the vehicles move along swiftly, safely and surely. (See Figure 12–9.)

Covered moving walkways transport pedestrians past buildings of strikingly new design within the city of tomorrow. Other highlights of the reborn city are high-speed, automatic vehicular parking, underground freight transportation, air shuttle service to satellite communities and rejuvenated urban living areas. The way of life has been made simpler and more pleasant for the motorist of the future visiting the heart of the city. The covered, moving sidewalks disperse pedestrians into shopping and business areas comfortably and easily. Parking facilities everywhere are automatic and plentiful. (See Figure 12–10.)

Energy from new sources will power the cities. Solar, geothermal and power beamed from geosynchronous, geostationary solar collector stations will orbit high above the earth's atmosphere. Garbage

Figure 12–9. The City of Tomorrow—a glittering complex of commerce and urban living, illustrating the use of air rights over existing thoroughfares. Low-rise dwelling units located above garage, commercial and recreational areas make up an inner-city community. (Photograph courtesy of General Motors Futurama Exhibit.)

Figure 12–10. Covered moving walkways transport pedestrians past buildings of strikingly new design within the "city of tomorrow." Other highlights of the reborn city are high-speed, automatic vehicular parking; underground freight transportation; air shuttle service to satellite communities; rejuvenated urban living areas. (Photograph courtesy of General Motors Futurama Exhibit.)

disposal problems will ease as city waste is converted into ethanol and methanol for either gasohol or catalytic conversion to gasoline.

"This last portal into the future concludes our trip of things to come and dreams as yet to be fulfilled. If you will now step before the next and last window you will be given a brief history and overview of this firm's goals and operation. In addition, 20 rear projection slides will be presented for your information and enjoyment. Each slide covers some major, interesting new design concept this firm has completed in recent years that in some way has advanced the state of the art or is on the cutting edge of technology."

"The firm is an engineering-architectural professional corporation, formed in 1985 and chartered in the District of Columbia, whose major services include consulting engineering and architecture. The firm specializes in structural, mechanical and electrical engineering with emphasis on energy conservation, value engineering, renovation and restoration of existing facilities, investigations, interior architecture and space planning. Firmly committed to the team concept, the goals of the firm are the pursuit of excellence in design, efficiency of production and maximum communication between the team members and the client."

"Three registered professional engineers direct the structural, mechanical and and electrical departments. Two registered architects are in charge of the architectural department. The balance of the firm is made up of engineering and architectural designers, draftspersons, space planners and administrative personnel."

"Recognizing the increasing importance of energy conservation, value engineering, and life cycle costing (LCC), the firm is organized with a strong emphasis on the engineering disciplines. When working as consulting engineers to architectural firms, the architects on our staff provide management services to facilitate coordination of the engineering disciplines, thus providing maximum communication between the architect and the engineers, and sensitivity to the needs of both the architect and the owner. For projects which stress engineering, the architects provide the support for design and integration of building systems."

"A design team, made up of the client, the architect, the project manager, the structural, mechanical and electrical engineers, and special consultants, works to develop the desired solution. In the process, alternative systems are compared. Those systems which meet the project requirements and are the most cost-effective in terms of life-cycle cost comparisons and energy efficiency are then integrated to give the most economical total design."

"Energy audits of existing facilities can uncover inefficient use of energy from oversized motors (overdesigned systems were the past standard practice), unnecessarily high illumination levels (requiring greater amounts of air conditioning), inefficient light sources, or air infiltration from windows and doors that are not weathertight."

"This concludes our presentation and we hope you have enjoyed it as much as we enjoyed preparing and presenting it to you. We hope you will come away from this audio-visual presentation with a better understanding of us and who we are, and what we are capable of doing for you—Beep—you have now returned to the present."

'Oh! Hi, John. I didn't see you standing there. I was so preoccupied with this very fine show you have prepared here. It is certainly good to see you in person after all the audio-video phone conversations we've had over the past few weeks.'

'Well, Paul, it's good to see you in the flesh, notwithstanding all the technological marvels of today that can bring one another into each other's home and place of business. Aside from our previous telex, facsimile transmissions and A/V phone conversations, is there any additional aspect of this firm I can tell you about or give you a deeper understanding of?'

'Not really, John. I feel as if I've always known you and this firm and know all there is to know about it. I am not sure if it's your personal charm and wit or the Public Relations abstracts you've written and sent me. As the partner in charge of my account, I feel my project couldn't be in better hands. I can readily see why they selected you as senior partner, John; you've got a way about yourself.'

'Well, in that case, Paul, would you like a tour of this rather unique building and, in particular, our office complex?'

'Yes, I certainly would, John!'

'As you may recall, Paul, this particular building was completed in 1987, just two years ago and we moved in just a little over a year and a half ago. I am sure you've already seen many of the new and innovative features as you came into and through the building proper. But let's amble through the various sections of the building and our office space in particular. I believe we have received all your initial design parameters by previous facsimile transmission and audio/video conferencing, and are currently working toward a 30% submission near the end of the month. Have you brought any additional data with you that you may wish to discuss or incorporate into the deisgn?'

'No, John, this is more of a social visit to your office and a somewhat unofficial walk-through of your premises to familiarize myself with your operation, office and personnel.'

'You may be noting, Paul, there are very few draftspeople in the office. It is not overly crowded by people. The reason is that all the unoccupied drafting tables of which there are only eight, are computerized automated drafting units with input data from floppy disc memory data storage, as well as graphic line trace technique equipment.

This not only reduces the personnel required and the time to produce a project but reduces the cost to the client. I might add that it not only reduces the time to spot design errors and mistakes in the dimensions strung out, but also catches any deviations in the various angles that are a part of every project. Quite simply, when a set of data, dimensions, angles, sizes and shapes don't add up to the configuration being laid out, the automated drafting stylus stops at the incongruent point where it no longer computes. This, as compared to a draftsperson who, in his earnest endeavor, attempts to make it work out and strings out two days of errors and misinformation that may take a design professional hours to trace and rectify.'

'Believe me, John, I know where you're at. I've had a few architects and engineers flog me with that syndrome.'

'Of course, we do retain some draftspeople, Paul, as well as computer and graphics design technicians. Here we have our computer-aided design (CAD) graphics and computer relay terminal design screens, computer aided—a very efficient department.'

'I'll say. Where do you guys dream all this up, John?'

'In some of our wildest, most productive dreams, Paul. The telephone communications you have been seeing throughout this office and the building is Dimension Service rather than Centrex, thus eliminating the need for cumbersome call directors and attendant wiring.

'As our very special executive visitor, it is my definite pleasure to escort you to the various areas of special interest to you. The word processing function has been reorganized so that all communications are video, and no paperwork is created except for the original. Word processing is central, but we have local communications stations variously located throughout the office. These enable convenient, instant use of this facility with editing and printing capabilities at these points. As you can see, here is a typical situation of an employee giving dictation over the telephone to a typist at word processing in which you will almost instantly see the rough draft of that letter printed out on the employee's desk computer relay terminal. This CRT gives the employee the capability of scanning the letter and making corrections with a magnetic marker. The corrected copy can then be stored in a magnetic memory file. Interoffice letters can and are transmitted to the other employees on their CRTs. They in turn may read it and decide to keep a hard copy printout from the CRT or release the image after reading it.

"Also, utilizing these CRTs at an employee's desk or in group areas, they can form a data communications link which instantaneously and company-wide, distributes related messages to employees, allowing hard copy printouts of these if required. It also provides access to an electronic library and can be used in preparing specifications. These communication units are served by an equally innovative and flexible underfloor system of branch circuit distribution from panelboards and telephone terminal key equipment which are located in central service closet clusters, and is the latest in cellular underfloor duct distribution systems. The underfloor ducts are provided with preset inserts at modular intervals of 5 feet on center which accept system outlets such as duplex convenience receptacles, telephone amphenols, CRTs and data system connectors. This enables us to have power and communications services available at regular modular intervals which offers us the utmost flexibility in space planning and work station locations and relocations. In fact we went one step further, and all conductors for receptacles and appliance branch circuits were specified and installed as minimum #10 AWG in lieu of the normal or conventional #12 AWG, in order to comply with NEC requirements for conductor derating to 60% when 28 conductors are run in a single raceway or cell compartment. Thus, a #10 wire which is normally rated at 30 amperes of current-carrying capacity is derated to 18 amperes at 60% derating. This just exceeds the continuous ampere rating of a 20-amp branch circuit breaker which is 16 amperes. Thus, the #10 conductors which are code derated to their new loading capacity exceed that required of the circuit breaker, while still maintaining a margin of capacity useful in counteracting voltage drop and excessive heat energy losses in the form of I^2R losses.

"It is envisioned that in order to accommodate greatly increased capacity requirements in the near future, the already large capacity underfloor ducts and trench header will have fiberoptic cables installed later next year in order to transmit video data and possible audio data if it should prove feasible. This fiberoptic system will increase the capacity of the underfloor duct system perhaps 1000-fold over a conventional present day system. Use of fiber optics will also eliminate the problems of fire danger due to equipment malfunction, accidental miswiring or electrical interference by power cables with communications signals. Fiberoptic cables can also be run alongside power conductors without separation or costly shielding.

"In addition, we have a very innovative energy conserving automated lighting control system being used in this office. It is called an automatic energy control system and works to maintain a given preselected conservative lighting level without undue waste of electric energy. In most office buildings, where a given 70 footcandle minimum level of lighting is desired, the lighting system is designed to produce an initial 100 fc. This design procedure is required in a conventional system since if the maintenance factor were not considered when establishing a maintained illumination level, a lighting installation designed for that particular fc level would start at that value only initially with the illumination level then steadily decreasing from the first day of installation. This is due to lamp lumen depreciation and lamp and luminaire dirt depreciation which together represent the maintenance factor.

"Thus, when using lamps and lighting fixtures rated at 0.7 maintenance factor, a typical lighting system intended to deliver 70 fc is designed to deliver an initial 100 fc. Over the life of the lamps, such a system will provide an average 70 fc but will provide an actual 70 fc for only part of its life. However, the energy consumed over the period of usage is 100% of the connected load whether it produces 100 fc, 70 fc or 50 fc.

"With an AEC system, available from several manufacturers, an initial illumination level of 100 fc is designed in order to maintain 70 fc at the end of the life cycle. Only 70% of the energy is utilized initially to produce 70% of the initial illumination, or 70 fc. As the system ages, more energy is required until many thousands of hours later, full 100% energy usage is required to maintain 70 fc. However, the difference in energy usage over those thousands of hours represents a substantial cost saving while maintaining the desired 70 fc throughout the installation's life.

"With AEC, fluorescent lamps can be utilized to the end of their life cycle (37,700 hours), while a conventional fluorescent lighting system would be relamped at 70% of maintained lumens (for relamping at approximately 16,000 to 18,000 hours). The AEC-controlled, fluorescent dimming system thus reduces the maintenance cost attributable to more frequent labor and material charges required by a conventional, non-dimmed, fluorescent lighting system.

"The AEC works by controlling an automatic dimming system with a photoelectric control which senses available light in the room. Set to sense 70 fc, the photoelectric control, in addition to reducing the output of a new (initial 100 fc available) system to 70 fc, will also take into account available daylight. Hence, if the daylight contribution to a room is 21 fc, the light output of the fluorescent fixtures, already operating at only 70% of full capacity, will then be reduced an additional 30%, providing a full maintained 70 fc at 49% energy consumption for the fluorescent system.

'Well, Paul, there you have it in a nutshell. Any questions?'

'John, you'll have to give me time to digest all this. Your organization is obviously tops. I've never seen anything like it. However, since this is just a stopover visit, I should be preparing for my trip to California. I'll have to be boarding my flight within two hours. How do I get from downtown Washington, DC to Dulles International Airport in the most expeditious manner?'

'Well, Paul, just walk 1½ blocks down to the blue line at North Farragut Square, corner of Conneticut and K Streets, N.W. Take the rapid transit subway system to the Union Station terminal, disembark and proceed to the top level of the parking structure in the rear. There you can get a shuttle flight from the heliport to the McLean heliport terminal. From there you proceed to an adjacent surface train station and ride the elevated monorail system out to Dulles Airport. The monorail glides right down the center of the previously installed divided highway's median strip.'

'It's been a pleasure talking with you, John, and getting to see you and your office. So I'll be saying good-bye for now.'

'Good-bye, and I hope to see you again soon.' "

ABBREVIATIONS

A area
AC alternating current
A.C. air conditioning
AEC automatic energy control
A.F.F. above finished floor
AWG American Wire Gauge
ASHRAE the American Society of Heating, Refrigerating, and Air-Conditioning Engineers, Inc.
BAS building automation systems
BEPS building energy performance standards
BRH Bureau of Radiological Health
BX General Electric's name for armored interlocked cable (also used generically)
CAD computer-aided design
CAD Computer aided drafting
CADD Computer-aided design and drafting
CESI Controlled Environment Systems Inc.
CFM cubit feet per minute
CM circular mils
CRI color rendering index
CRT computer relay terminal
CRT cathode ray tube
DC direct current
E.A.S.E. Energy Activation Switching Equipment
ECALO Energy conserving automatic light output
EMF electromotive force
EFF efficiency
EMI electromagnetic interference
EMT electrical metallic tubing
ER ellipsoidal reflector
ESI equivalent sphere illumination
F frequency
fc footcandle
FCAN full capacity above normal
FCBN full capacity below normal
GSA General Services Administration
HID high intensity discharge
hp horsepower
HPS high pressure sodium
HVAC heating, ventilating and air conditioning
I current
ID inside diameter
IES Illuminating Engineering Society
IR infrared
JB junction boxes
kVA kilovoltamps
kVAR kilovoltamps-reactive
kW kilowatts
kWh kilowatthours
L length
LCC life cycle costing
LDD lamp (luminaire) dirt depreciation
L.E.S.S. Light and Energy Saving System
LLD lamp lumen depreciation
lm/W lumens per watt
LPS low pressure sodium
M meter
MCB main circuit breaker
MF maintenance factor
MH metal halide
M.H. mounting height
MHD magnetohydrodynamic generator
MLO main lugs only
MV mercury vapor
NEC National Electrical Code

NEMA National Electrical Manufacturer's Association
OC on center
OEM original equipment manufacturer
O.S.H.A. Occupational Safety and Health Act
P power
PF power factor
R resistance
RFI radio frequency interference
RIF resource impact factor
rpm revolutions per minute
RS rapid starter
RUF resource utilization factor
S/MH spacing to mounting height
SMH super metal halide
SPD spectral power distribution
TSCA Toxic Substances Control Act
T.H.W. (type of conductor insulation)
UL Underwriters Laboratory
UPS uninterruptible power source
UV ultraviolet
V voltage
VCP visual comfort probability
VDT visual display terminals
VHO very high output
VLH very low heat
W watt
X reactance
X_C capacitive reactance
X_L inductive reactance
Z impedance
ρ resistivity
ϕ phase
μ micro
θ theta

INDEX